March 6, 2004

Dear Woodie,

W[illegible] regards

— a[illegible]n

your pioneering

work

[illegible]

Religion and the Formation of Taiwanese Identities

Religion and the Formation of Taiwanese Identities

EDITED BY
Paul R. Katz and Murray A. Rubinstein

palgrave
macmillan

RELIGION AND THE FORMATION OF TAIWANESE IDENTITIES

First published 2003 by
PALGRAVE MACMILLAN™
175 Fifth Avenue, New York, N.Y. 10010 and
Houndmills, Basingstoke, Hampshire, England RG21 6XS.
Companies and representatives throughout the world.

PALGRAVE MACMILLAN is the global academic imprint of the Palgrave Macmillan division of St. Martin's Press, LLC and of Palgrave Macmillan Ltd. Macmillan® is a registered trademark in the United States, United Kingdom and other countries. Palgrave is a registered trademark in the European Union and other countries.

ISBN 0–312–23969-6 hardback

Library of Congress Cataloging-in-Publication Data
Religion and the formation of Taiwanese identity/edited by Paul R. Katz and Murray A. Rubinstein.
p. cm.
Includes bibliographical references.
ISBN 0–312–23969–6
1. Religion and politics—Taiwan. 2. Nationalism—Taiwan. I. Katz, Paul R., 1961– II. Rubinstein, Murray A., 1942–
BL65.P7 R43355 2003
200'.95124'9—dc21 2002033306

A catalogue record for this book is available from the British Library.

Design by Newgen Imaging Systems (P) Ltd., Chennai, India.

First edition: April, 2003
10 9 8 7 6 5 4 3 2 1

Printed in the United States of America.

Contents

Chapter 1
The Many Meanings of Identity: An Introduction

Paul R. Katz and Murray A. Rubinstein

This volume centers on the creation of varied forms of individual and group identity in Taiwan, and the relationship between these forms of identity—both individual and collectively—and patterns of Taiwanese religion, politics, and culture. We explore the Taiwanese people's sense of who they are, attempting to discern how they identify themselves—as individuals and as collectivities—and then try to determine the identity/roles individuals and groups construct for themselves. We also explore how such identities/roles are played out within the family and peer group, at the local level of the village, town and neighborhood, and on the regional level, the national level, and within the larger Chinese cultural/religious universe. In this volume, we seek to answer questions about the complex nature of identity/role and the processes of identity formation, and then determine how such identities/roles are reflected in the religious, sociocultural, and ethno-political actions and structures that have shaped Taiwan's multileveled past and its many faceted present. In this introduction, we first suggest how individuals and groups in the overlapping realms of Taiwanese politics, religion, and cross-strait relations can utilize identities/roles as cultural constructs. We then define the concept of identity/role and the processes of identity/role formation, concepts and processes that, in Wagnerian fashion, act as a *leitmotif* that works through and resonates in different ways in the contributions to this book.

Religion, Politics, and Identity: The Ma-tsu Pilgrimage to China

One recent example of the intense and also complex links between religion, politics, and identity in contemporary Taiwan involves the attempt by one of Taiwan's leading Ma-tsu temples, the Chen-lan Kung of Ta-chia (Taichung County), to undertake a direct pilgrimage to the cult's ancestral temple (*tsu-miao*) in Meichou (Fukien).[1] When Taiwan and China established informal contacts back in 1987, people from Taiwan who rushed across the Taiwan Straits included war veterans who had accompanied Chiang Kai-shek from China to Taiwan after the fall of the Republic of China in 1949, businessmen and industrialists seeking a chance to make a profit, and pilgrims seeking their religious roots. One of the first pilgrimages to China was undertaken by the Chen-lan Kung that took advantage of the occasion to assert its legitimacy and authority (or "incense power") as one of Taiwan's leading Ma-tsu temples.[2] Between 1987 and 2000, numerous temples organized pilgrimages to China, but due to restrictions on cross-strait contacts were always forced to take indirect routes through other countries such as Hong Kong. More recently, Chinese worshippers and officials also brought a statue of Ma-tsu from Meichou to Taiwan (via Hong Kong) for a pilgrimage of their own in 1998, a move which caused great controversy among those Taiwanese Ma-tsu temples that were willing to welcome this "Chinese Ma-tsu" and those temples to Ma-tsu that were not.

During the Winter of 2000, however, some began planning to try and challenge the status quo. The leadership of the Chen-lan Kung, which included the Speaker of the Taichung County Council Yen Ch'ing-piao, set in motion an effort to organize a direct pilgrimage from Tachia to Meichou. It was around this time that Yen attempted to blend the roles of local elite, religious leader, and political kingmaker, by declaring his support for the presidential candidacy of James Soong (Sung Ch'u-yü). Thus, a newspaper article published on February 18, 2000, reported that as early as one month before the actual presidential election took place, the Chen-lan Kung's leadership were anticipating difficulties should "one unnamed candidate" (the Democratic Progressive Party (DPP) nominee Chen Shui-bian) be elected.[3] As we will see later, this view proved to be highly realistic.

The issue of direct pilgrimage to China became more pressing during the spring and early summer, with people throughout Taiwan beginning to discuss the importance of such an event. Some enterprising local officials in the offshore islands of P'eng-hu (the Pescadores) and Chin-men (Quemoy) contacted members of the Chen-lan Kung temple committee and attempted to persuade them to route the pilgrimage through their counties (in order to profit from an anticipated influx of thousands of pilgrims), but to no avail. On June 4, Yen Ch'ing-piao and Kuomintang (KMT) members of the Legislative Yuan and National Assembly from central Taiwan gathered at the Chen-lan Kung to throw divination blocks in order to determine a date for the pilgrimage. The ritual proceeded smoothly, with the goddess setting July 16 as the departure date. From this point on, a political tug-of-war ensued between Chen Shui-bian's new government (which tends to oppose reunification with China) and members of the opposition, including elected representatives of the KMT, the New Party, and Sung Ch'u-yü's newly formed People First Party (PFP). Leading members of the DPP, such as Vice-President Annette Lü (Lü Hsiu-lien), who attacked the Chen-lan Kung temple committee for attempting to play politics with a religious event, while members of the opposition strove to pass resolutions designed to force Chen Shui-bian's government to approve a direct pilgrimage to Mei-chou. At the same time, Chinese officials, particularly local officials from Fukien who like their counterparts in P'eng-hu and Chin-men anticipated a windfall yet were also well aware of the symbolic importance of Taiwanese pilgrims traveling directly to China, made every effort to express their willingness to allow a direct pilgrimage, albeit with some strings attached (for example, only vessels from China or Hong Kong could be used to transport pilgrims). Caught in the middle were Ts'ai Ying-wen and members of the government's Mainland Affairs Council, who pointed out that regulations governing direct links between Taiwan and China had not been passed, and that there was no way to guarantee the safety or rights of those pilgrims who participated in this event. As for Chen Shui-bian, he attempted to persuade the Taiwanese people that going on a direct pilgrimage to China would only serve China's propaganda efforts.

On June 10, Tseng Ts'ai Mei-tso, who played the dual roles of KMT legislator and the chairperson of another leading Taiwanese temple, the Ch'ao-t'ien Kung of Peikang (in Yunlin County), claimed that Ma-tsu had appeared to her in a dream and told her to break the ongoing impasse between the Chen-lan Kung and the government by undertaking an indirect pilgrimage to Meichou. Views of why the Ch'ao-t'ien Kung chose this course of action varied. On the one hand, it appeared to represent a statement of loyalty (or at least sympathy) toward the new government; on the other hand, the Ch'ao-t'ien Kung had been the original destination of Tachia's Ma-tsu pilgrimage before 1987, and a game of one-upmanship may have been involved. Whatever the case may be, the government soon expressed its support for the Ch'ao-t'ien Kung's decision. As early July arrived, Yen Ch'ing-piao and his supporters realized that a direct pilgrimage was not feasible, and decided to follow the Ch'ao-t'ien Kung's example of taking the usual indirect route.

Leaders of the DPP and Chen Shui-bian's government attempted to placate the supporters of both temples, although it is worth noting that Chen chose to personally visit only the Ch'ao-t'ien Kung. On July 16, approximately four thousand pilgrims flew from Taiwan to Hong Kong, and from there headed to Mei-chou. The airlines, as usual, had to find a middle ground between safety and the sacred. For example, Ta-chia's Ma-tsu statues were not required to be X-rayed, but they did have to wear seatbelts on takeoff and landing. The pilgrims reached Mei-chou by July 18, with Yen Ch'ing-piao assuming the dual roles of cult leader and devout worshipper by helping carry the goddess' palanquin. The pilgrimage rites at the temple were noteworthy for a number of miraculous events. For example, when Ma-tsu's palanquin approached the Mei-chou temple, a rainstorm occurred, but quickly stopped once she had entered the temple's main hall. Members of both temples' leaderships declared the event a success.

Theoretical Issues

The above description of the Ma-tsu pilgrimage reveals that religion and religious events constitute key arenas that members of different interest groups attempted to utilize to achieve both economic and political goals. Members of the DPP, who have long advocated independence or at least opposed unification with China, saw Ta-chia's attempt to undertake a direct pilgrimage to China as a potential first step down a dangerous path. Members of the opposition were less fearful of such a result, and also used the entire affair to attack Chen Shui-bian's new government. The events described above clearly reveal the importance of identity, as well as the different roles (politician, temple committee chairperson, devout worshipper, and the like) people play in attempting to create or reinforce differing identities. In the pages that follow we attempt to elucidate the concept of identity and its multiple levels of meaning. In doing so, we argue that to understand what identity is we must link it to a related construct: role. We argue that on the most basic levels of psychosocial organization—the levels of the individual, the self, and the group—identity and role are tied by perception and self-perception (or consciousness) and by praxis. We then trace the nature of these linked concepts of identity and role as they took form in the Western and Chinese thought. In doing so we enter the realms of psychology, sociology, philosophy, and religion as we attempt to grasp the often unstated—or subtextual—nature of Western and Chinese approaches to questions of identity, identity/role, and identity/role formation. We conclude by showing sets of linkages between these constructs and modes of praxis and patterns of religion—and religiously related cultural practices—that are found on the island of Taiwan.[4]

Before we begin, however, it will be necessary to first confront three central questions that drive this introduction and this book. What is "identity"? What are the levels of meaning captured by the term "identity"? What is the relationship between the concepts of "identity" and "role"?

The first question is at once the easiest and hardest to answer. Identity, even more than role, has been a subject that philosophers, psychologists, sociologists, political scientists, and historians have all dealt with in a various ways that usually reflect the contours of their fields or subfields. The term "identity" can be defined in a number of ways. In the *Webster's Collegiate Dictionary* (5th edition) the definitions of the term are as follows: (1) Sameness of essential character; sameness that constitutes the objective reality of a thing; (2) Unity and persistence of personality; the condition of something described or asserted. Since we are dealing with identity as a psychological, psychosocial, and social construct it is the second definition that is of most value to us.

The term, of course, is related to a similar term, identify, which is defined by Webster as: (1) To make to be the same as; to consider as the same in any relation, and (2) To prove the same (with something being described, claimed, or asserted). This linkage of terms is essential. Identity is thus a steady state—a manifest psychosocial/social reality—while identify suggests process or praxis. A quick search of various search engines on the World Wide Web provides a sample of the various literatures that have evolved on the subject of identity and on the related subject of identity in its relationship to ethnicity as well as geographic entities such as county, province, region, and nation.[5]

There are also levels of the term "identity" as we use it here. There is the realm of the individual. Here one defines over time one's own identity—one's own complex nature. Identity here implies an organism—a sense of wholeness: one has a distinctive identity or personality. But the term can also be parsed to suggest that one can identify with various sets of feelings or beliefs that are held by others. When used this way identity becomes both a psychological and a social construct. One sees oneself as identifying with any number of formal organizations and/or local, regional, or national units. By doing so one assumes within oneself an array of identities.

Let us recapitulate, for a moment: We use the term "identity" in two distinct ways. First to suggest both a personal working out of questions of "who am I"—a search for the essential self or core role—and also as an outward gazing social component of the personality that fights against the tendency toward solipsism. However, any individual's life is defined not only by who he or she is—the core self-vision we term identity—but also by what he or she does. One is said to be playing a variety of different roles, roles that are assumed when one is with one's nuclear family, when one is with ones extended family, when one is with one's friends, when one is with members of one's occupational peers, and one is involved in activities in the larger society. Thus, a second facet of identity—as component—is created by the act of identifying and feeling that one is part of the community or sets of communities that include family, local society—that is, neighborhood groups, occupational groups, the immediate sociopolitical group—the club or local party headquarters, or a spiritual–cultural body or bodies such as sects, temples, or churches. Assuming any such social identity or collectivity of social identities makes one feel part of a greater social or sociopolitical or spiritual–culture whole—that is, a community. In some cases, these feelings dominate a person to the point that the role that evolves is the one an individual most frequently assumes or assumes regularly at times in the political or the socioreligious year.

The construct "role" is closely linked to but not exactly the same as the construct "identity." Identity here is what one feels—how one sees one's self. Role is what one does. Identity and role are constructs that are inextricably tied together. They form, to use a classic Chinese framework, a *nei* (inner)–*wai* (outer) dyad basic to any individual. Identity here is seen as a psychological construct. One's identity is what one is, but one is also what one does. Thus, we can see the linkage between the two. As I have also suggested, just as one can also identify with many other groups or social bodies one can also assume roles that reflect such identification processes—that sense of membership with different communities that are arrayed about one is a set of concentric circles.

I have laid out the clear links between identity and role. Here, let me deal with the deeper nature of the construct "role." In the West, "role theory" begins, it is often said, with a theatrical model that is used to explain the development of personality and behavior in the context of a given individual's "concentric circle" of sociocultural relations. The oft-noted starting point for many role theorists is the Duke's monologue in *As You Like It*. "All the world's a stage," Shakespeare's Duke tells us, "and all the men and women merely players. They have their exits and entrances and one man in his time plays many parts."[6] Role theorists use this idea of stages as a foundation for more elaborate descriptive theories.[7]

Theories of role and role formation vary but share one common and often criticized characteristic—they are often descriptive and not analytical in nature. This characteristic (or "fatal flaw" some might say) makes such theories useful to someone describing a given structure of personality but not to one probing deeper and analyzing that structure—of that moment of role-driven behavior. One such theory, which I find most appealing, begins with the idea that there exists a core self—or identity—created by a set of genetic and environmental factors in a person's life. This core self/identity remains as the center of one's being. As one develops and matures one sees how one must fit into the larger social environment, and one must thus adapt and play a given role or sets of roles. And as one passes through life one may assume roles appropriate to that given stage of one's own existence. Age, experience, and knowledge combine to help a person create different role sets at each new stage in his or her own life. One common model, developed by a sociologist in the United Kingdom, one I feel is the best fit here, is to see the core role as a tree trunk. As a tree grows, its branches sprout and mature and change; but its trunk, albeit somewhat larger in girth, remains much the same, providing the necessary solid base and system of roots for support over time. Those scholars who have worked on role theory have developed a set of descriptive constructs that can be useful in allowing us to look at a person's life even as one factors in the distinctive "stages of life" as the individual develops moment by moment, adjusting to given environments and circumstances and to the realities of ever-advancing age. Such theories begin with certain basic assumptions. First, personal change is guided by a person's experience of action in roles that the individual undertakes. Second, human behavior can be understood as an interaction between their enduring characteristics (personality or identity) and their social situations (roles). Third, social interactions can be, to a fair degree, understood in terms of the positions that actors occupy in a given social structure, as well as the ways in which their perceptions are molded by the very values, sanctions, and perceptions they have of the given roles or sets of roles they occupy. Finally, and for our purposes this a key fundamental element in role—"the larger social system can be understood in part as a network of positions and associated behaviors."[8]

One can see that roles are not simply static and that roles and sets of roles evolve as an individual plays them out over his or her lifetime. Role, thus, consists of both prescribed structures and spontaneous innovations. Furthermore, role-playing is defined by the individual as well as the organization—family, occupational group, recreational club or society, and social network—that the individual belongs to. Thus, there are sets of role expectations and role enactments. There is also a degree of individual involvement in a given role. Finally, there are problems associated with roles and the individual's assumption of role. One takes on a number of roles on any given day and in any given stage of life entailing a constant transition from one role to another. The assumption of roles leads to distinct problems of expectations and the ability (or lack thereof) to fulfill the expectations of a role or set of roles. There are thus issues of "role quality" as viewed from the outside—from the viewpoint of one or more individuals who view that actor/individual—and the inside: that specific individual's sense of "how he is doing."[9]

In this volume, the concepts of role and identity are used in a variety of ways. We first look at the process of identity formation over time. Chang Mau-kuei examines this process in chapter 2. As we read his chapter on the origins and transformation of Taiwanese National Identity we find that the functional aspects of identity and role as well as the nature of role as a means of defining one's self within regional and provincial environments in evidence even as identity formation takes place. Who one is and what one is are subtexts in this powerful essay. Duujian Tsai deals with similar issues in chapter 3, buts sees them as state of consciousness and not as historical process. He also shows how studies of ethnicity and identity help define the popular as well as intellectual discourse. Su-mei Wu's chapter (chapter 4) shifts to a related set of issues—the way specific forms of cultural performance help to define both individual role and two levels of identity—individual and regional/provincial. Michael Rudolph's chapter (chapter 5) on the *yuanzhumin* (the aboriginal tribal peoples who were the first settlers of the island of Taiwan) focuses directly on identity that is both individual and aggregate.

Those chapters that focus on religion also show the clear relationship between religious development and different forms and levels of identity and role. Paul R. Katz spells out the nature of that relationship and its complex dynamics in chapter 6. In chapter 7, Murray A. Rubinstein looks at identity as corporate but also spells out the nature of functional roles in his study the medium as message paradigm as it functions in temple communities' modes of self-presentation. In chapter 8, Avron A. Boretz provides an example of identity/role in his study of martial arts troops that take part in village wide religious rituals and cultural celebrations. Finally, in chapter 9, P. Steven Sangren pulls these threads together and deals with levels of identity and their relationship to religion as seen in modern day Taiwan.

Identity/Role in Chinese Culture

We have looked at both identity and role in the same way as Western thinkers and theorists. Now we must try to understand how Chinese thinkers and those involved in Chinese sociopolitical and religious reality conceive of role and identity. Perhaps the best source for a classical Chinese perspective on identity, role and different "roles," as well as the necessity and consequences of assuming certain identities and of playing such roles is the first chapter of the *Ta Hsüeh* (The Great Learning). This text, the brilliantly conceived chapter in one of the two smaller classics that make up the core Confucian canon, known as the *Ssu Shu* (the Four Books), defines the idea of what patterns of roles individuals must learn to play if they are to be true to the Confucian *Tao* (the Way). In a series of carefully constructed linked clauses, the author of the *Ta Hsüeh* lays out what series of roles the gentlemen—the *shih* and the *chün-tzu*—must assume in each carefully described stage. We must add that this passage also describes the model of the core identity of the *ju* (or Confucian master)—or would-be *ju*. In this scenario we have a person moving—in his mind at least—from the outer realm (*t'ien-hsia*, literally "all under heaven") to the inner realm and then to the outer realm once again. This links the individual and his pursuit of knowledge to the world system. One knows the meaning of things—*ko-wu*—and to implement this knowledge is the basic task of the scholar–official. His role is central to the regulation of the family, the clan, the nation, and the world. Here, we have a melding of identity—as public statement and role—as the active component or the agent of praxis that makes the Tao real in the daily world of the *shih*. To be sure, not all men can play this role (despite the fact that the Confucian canon contains a model of social structure that clearly defines roles and occupations). Furthermore, we see relationships and roles spelled out in the five core relationships of Confucianism that further explicate what individuals must do to serve the larger society they are part of.[10]

There are of course alternative concepts of roles, as varied as the Hundred Schools that grew out of the Axial Age. The Axial Age was that period in the history of the classical—and pre-Christian—civilizations when these foundational civilizations developed their first sets of "formative world views and philosophies." The members of the school who wrote the *Ta Hsüeh* are known as the *ju*—or to use the Latinate term the Jesuits introduced—the Confucians. The other philosophical school that stands most clearly opposed to the theory of social and familial responsibility laid out in the *Ta Hsüeh* is the Taoist School.

Taoist views of identity and role are spelled out first in the *Tao-Te Ching* and *Chuang Tzu*. Related theories and patterns of Taoist identities and roles also took shape in proto-Taoist religion, especially the arts of the arcane and the supernatural as practiced by early specialists known as "masters of techniques" (*fang-shih*). Moreover, cosmology and correlative theories produced during the period, dating from the completion of the Huai-nan Tzu to the core centuries after the fall of the Han, helped shape the thought of those Taoist masters whose works and revelations make up the core of the early Taoist Canon.[11]

The Taoist concepts of identity and roles that we find are first suggested in the various chapters of the *Tao Te Ching*. Here, we find the ruler who rules by not ruling as an exemplar, and the man who has nothing but his freedom as another. In the *Chuang Tzu*, we have the master himself playing the wise fool in his conversations with the powerful. We also have men like the sheep butcher who teaches us that in time and through mastery he learns how to let the blade guide itself. The butcher plays a specific role but also makes it an intrinsic part of his inner self—it becomes his identity by allowing this degree of integration to take place as he unites with the *tao*: that is, the essence of the universe. This idea of a learned and internalized mastery as a goal for the average man is an important one that allows the Taoists to differentiate themselves from the status and formal set of roles—as advisors and court scholars—so essential to the Confucian school.[12]

Buddhism's entry into China and its growth during centuries of disunion led to the development of new and more complex sets of religiously centered roles. Buddhism itself does not have a formal theory of identity and role but those who developed this philosophical/faith system recognized the need for a structure for the *sangha* (community: that is, the monastery/clergy) and for rules to define the operation of the Buddhist monasteries and centers. This in turn produced a whole set of new occupational categories that are dealt with in that part of the Buddhist cannon that covers community and religious organization. Yet, there is a contradiction here—one made necessary by the pragmatic realties of running a religious establishment: Buddhists preach a doctrine of nonexistence and nonpersonality—or, to use our terms, nonidentity. However, to keep their faith alive Buddhist leaders created both clear-cut inner identities as well as role sets and hierarchies of organization that place more people in the world even if it is a Buddhist centered world. The contradiction is accepted because it makes the Buddhist presence—a presence that is designed to allow teachers and missionaries to preach the message of an end to existence—a viable one in our complex world.[13]

The Taoists responded to the formation of a Buddhist religious establishment by creating one of their own. They also developed a whole set of formal religious doctrines, teaching, and revelations that are compiled in the Taoist Canon. These Taoists also founded religious centers and formal abodes where doctrine was taught and rituals performed, and where the composition of new scriptures could be undertaken. The way was thus opened for further occupational specialization. Taoists began to play roles in local community life and became specialists who performed important rituals at designated periods every few years. The religion of natural force and of an ever-shifting reality thus became, like the Buddhist faith, one tied to the material realities of the world and the necessity of creating, running, and supporting large-scale organizations that allowed the faith to sustain itself and grow.[14]

Popular religious organizations, as well as more esoteric and sectarian cults, have also created sets of roles for their believers, modeled upon those roles and occupations yet different from those one could see within the Buddhist and Taoist traditions. While popular cults existed in ancient times it was during the T'ang and more particularly during the Sung Dynasties that specific cults arose. This era also witnessed the development of pilgrimage networks consisted of leading sacred sites and branch temples, often linked by division of incense (*fen-hsiang*) rituals.[15]

The temple cults referred to above were complex socioreligious entities that needed dedicated staff, temple committees, and caretakers in order to function. Such temples also needed a supply of statues of their deities—both large and small—and this in turn provided occupations and roles for those working with or whose livelihood depended on the temple. In addition, a wide range of religious specialists could be directly or indirectly affiliated with temples, including Taoist priests, shamans, and spirit mediums, and this further contributed to the creation of new religious roles. While there are sets of what I would call temple gazetteers, there does not seem to be a formal set of works about religious practitioners or for religious practitioners. To get some sense of roles and organizations, one must visit modern sacred sites on Taiwan and on the Chinese mainland, and get to know the people who work there as well as obtain books and pamphlets published by that temple or group of temples.[16]

What we have tried to suggest, albeit in a rather cursory fashion, is the way the major canonical and non- (or quasi) canonical religious traditions in China defined and developed roles for their believers and potential leaders. The roles defined are of those who would be exemplars, leading others or mediating between others—the average man or woman—and the forces of the supernatural that make demands upon those who inhabit our own realm of reality. Let us take this argument a bit further and amplify these comments on role and role models in the popular religious tradition (*min-chien tsung-chiao*). In this tradition, we find identity and role performance at the very center of things. The gods who inhabit the popular pantheon in China and Taiwan usually began as human beings—historical personages. One famous exception to the rule is the God of Literature (Wen Chang) who began as a great serpent, as Terry Kleeman has shown us in his valuable study.[17] After they died they continued to serve the people, and the most well known were then recognized by local governments and perhaps even by the imperial court. The great gods and goddesses were thus exemplars that one could revere and worship. Large-scale cults developed around these gods. Kuan Ti is perhaps the most famous throughout all of China but Kuan-yin, Ma-tsu, Pao-sheng Ta-ti, and the Royal Lords (*wang-yeh*) are all important deities in local villages in specific counties, in certain provinces, and sometimes in entire regions, like China's long eastern coast. These gods have networks of temples and, as I have suggested, it is at these temples that people play specific roles designed to serve the needs of the god or the local community of believers, as well as the larger community of pilgrims who may visit that temple site—as is the case in Pei-kang and Hsin-kang on Taiwan and at Meichou a small, and now often-visited island that lies a few miles east of (and within the borders of) Putien County, Fukien.[18] The popular tradition thus sparks the creation of a set of role models—for a temple-centered/cult. Sects such as the Unity Sect (I Kuan Tao) on Taiwan do much the same thing, working to attract a core of believers and then creating formal roles for those within the sect—the high-level executives, middle managers, and the office staff—to meet the needs of the expanded temple/sect communities.[19]

Identity/Role in Modern Taiwan

Now let us return to the concept of identity. Role, as we suggested earlier, can be seen as an intrinsic aspect of identity. One takes and plays a role, and assumes a given identity. If that identity is one that is connected with one's core self—or core role—then that sense of identity is one that transcends the person—that places him with others. It is at that moment in time—the narrative—that identity becomes a bridge between person and community. Now we can go a step further. When one joins with many in accepting a given identity within a religion, or as a member of a religion, or a citizen loyal to one specific community or region or province or political body, then one becomes a part of community of shared belief. One who is now a part of this collectivity can act with the greater strength that sheer numbers convey. Such an "identity community" or set of identity communities are what we find when we examine the processes of ethnic-centered or faith-centered politics in the West and in Asia. Over the past few decades, political scientists, as well as those who study the sociology of politics and anthropology of politics, have begun to examine and analyze the ways in which individuals play out their roles as members of specific communities.[20]

One must be clear that the individual who belongs to such communities may also join other similar communities based upon different types of identities. They also join because of the opportunity to play different sets of roles and thus bond with different role-based associations. One plays many roles—and assumes different identities or sets of identities. Many of these roles reinforce specific types of a person's subidentities—those facets of one's self that center around "belonging" to a given group or sharing goals, sympathies, and life style with its members. This structure of identification-based identity/role is what the author of the *Ta Hsüeh* was thinking about in his first chapter. What that ancient author did, but what those acting out such roles today may not do, is to make the kind of linkages that created the double spiral found in this Confucian classic.

If we examine modern Taiwan we see that such patterns can be found. One moves from roles in the collectivity of the family to roles in the place of employment, to roles in the realm of religious life, to roles as members of a given peer recreational or cultural performance group. Each body or collectivity makes its own demands on a person, and a person who wishes to fit in will accept these demands and act accordingly. One, thus, transforms one's identity by adding on subidentities and identifying with different sociocultural clusters and groups.[21]

Religion is a part of this process, as I suggested, in a number of ways. First, it provides another specific collectivity one can identify with. Second, it may provide functional roles that satisfy an individual's core identity or self. Third, it may convey status both by simple membership and by providing distinct roles to play at given moments in the year. Finally, it may provide a sense of inner strength that enables an individual to cope with his or her ever more complex life in an increasingly complex and postmodern Taiwan. Chang Mau-kuei's essay, chapter 2 in this volume, examines the evolution of identity in Taiwan since the

Japanese period and deals with the impact of the Taiwanese' sense of identity in contemporary Taiwan.

But what of the deeper links between notions of what I can now term identity/role (and the dynamics of identity/role formation), the creation of distinctive sets of identification-based subidentities and religion? The authors who have contributed to this volume have attempted to draw on a sizable body of scholarship pertaining to the links between religion, culture, and identity formation. One of the leading scholars in this area is the sociologist Hans Mol, who in a pathbreaking study he wrote twenty years ago, argued that religion represents the "sacralization of identity."[22]

The work of Mol and other social scientists has proven highly relevant for scholars doing research on both the historical growth of Chinese religions or their current development in Taiwan. For example, Mol and his colleagues have shown that the creation of identities by means of religious activities is often linked to responses to drastic socioeconomic and political changes.[23]

This should come as no surprise to Chinese historians who have studied the growth of Buddhism and Taoism during China's tumultuous medieval era, the rise of the new Taoist movements in North China following the fall of the Northern Sung Dynasty (960–1125), the development of popular religion in the Southern Sung (1127–1279), and the increases in activity on the part of organized religions during the waning years of the Yuan (1279–1368), the Ming (1368–1664), and Ch'ing Dynasties (1644–1911).[24] Social scientific research on religion and identity also reveals the relatively conservative nature of identity formation, especially how such processes tend to result in the preservation, restoration, or reinforcement of already extant identities, as opposed to the formation of new or original identities.[25] As we will see later, the data presented by the contributors to this volume indicates that most forms of identity expressed in Taiwanese religious activities tend to be conservative in nature.

In terms of the actual processes by which collective identities may be formed or reinforced, Mol postulates the existence of the following four mechanisms: (1) Objectification, defined as the "projections of order in a beyond where it is less vulnerable to contradictions, exceptions and contingencies"; (2) Commitment, that is, the "emotional attachment to a specific focus of identity"; (3) Ritual, especially ritual which "restores identity. . . when disruption has occurred"; and (4) Myth, that is, narrative accounts that "provide the fitting contour for one's existence, sublimating the conflicts and reinforcing personal and social identity."[26]

The chapters in this volume, confirm the importance of all four mechanisms in various processes of identity formation, although in many cases ritual's role appears to outweigh that of the other three as demonstrated in the chapters by Avron A. Boretz, P. Stephen Sangren, and Su-mei Wu.

The Structure of the Volume

The essays can be thought of as forming a conversation or, more accurately, a series of related and often overlapping conversations. Conversations are often not neat and self-contained. They are messy space/time events that flow and ebb and change given the nature of the individual then taking center stage. Thus, the reader will find, as did the scholars who reviewed this manuscript that the editors have put together, that the ideas contained in some of the essays seem directly opposed to those in others. Approaches, methodologies, and conclusions seem wildly divided even though the ostensible subject matter seems much the same. There is an openness that, the editors believe, is very much a matter of serendipity rather than formal design. Had the editors commissioned the articles, as did Rubinstein in his previous edited volume, *Taiwan a New History*,[27] rather than gathered them together, the flavor of the book would have been very different. The book before you would be, no doubt, more coherent perhaps, but less alive with the sense of dialogue and intellectual exchange. To state what will be soon clear to you the reader, no rigid ideological or intellectual line has been imposed on the authors, nor have they been instructed to make specific points or adopt specific methodologies or modes of analysis. Each author speaks with his or her own distinctive voice and writes in his or her own formal discipline—or, as some do—in a determinedly interdisciplinary fashion.

At the same time, however, there is common ground in the larger sense that all authors deal with issues of identity and identity formation. They do so in different ways and with different sets of objectives. Some deal with identity as *ding-an-sicht*—as an entity or construct in-and-of itself. Others see identity within a cluster of related constructs such as role. Still others see identity in terms of a force that creates ethnicity or nationalism, as well as local and institutional bonds and loyalties. Furthermore, some authors see identity within a wide range of contexts, and thus focus upon identity and the problems it often creates as expressed in the functioning of society and in the socially constituted realms of politics, culture, and religion.

While there is free flow there is no chaos. There is a method of sorts to be found in the structure—here meaning order in which the chapters appear—that the editors have created. For example, chapter 2 in this collection, Chang Mau-kuei's, has a long narrative line tracing the history of the issue of identity in Taiwan: it serves to set the framework of the discussion that then takes place. The last chapter in the collection, P. Steven Sangren's, serves to tie up various threads even as it broadens the discussion in terms of both space and time. These two strong chapters, both of which feature large-scale theoretical constructs and solid data serve as two bookends, firmly anchoring the discussions in time and in space. Other chapters such as Tsai's and Katz's, are designed to introduced more specific sets of theoretical constructs that deal with issues related to identity—as it is influenced by and reacts on religious consciousness and religious development. Finally, the chapters by Rudolph, Wu, Rubinstein, and Boretz provide both specific modes of analysis and detailed data that strengthen not only their own individual theses but also positions that other scholars have taken.

Let us now introduce our authors—and their individual statements—in this dialogue-from-a-distance that you the reader partake of as you read our book.

Chang Mau-kuei leads off with his overview of ethnicity and ethnic conflict on Taiwan. He gives us a detailed look at the evolution of the issues of ethnic and national identity as they are now defined in Taiwan. He thus provides the historical context for many of the other chapters that follow. In effect, he begins what will evolve in this volume—a multivocal conversation on the issues of identity formation, role and identity definition, and ethnic conflict on the small but very complex island that is Taiwan.

Tsai Duujian provides another perspective on issues of identity by examining these issues in what the editors see as a dense, complex, and often-provocative essay on the now very "hot" questions of national and regional self-definition. He begins with a discussion of the work of Alan Wachman, using his *National Identity and Democratization*[28] as a sort of straw man to present Taiwanese perspectives on issues of ethnicity as well as conflicts on the island. In doing so, he attempts to develop a corrective to Wachman's well known and respected take on these issues.

The linkages between specific forms of cultural performance and issues of cultural and ethnic identity are spelled out by Su-mei Wu in chapter 5. In this chapter, Wu examines the traditional art form of hand puppetry. She first gives the reader a history of the art form and explores how it has changed during Taiwan's modernization. She also examines the ways in which the practitioners of this art have adapted to the realities of the new forms of media found on the island, most notably television.

Chapter 5 by Michael Rudolph is taken from his doctoral dissertation. It examines the attempts of scholars among Taiwan's aborigines (*yuan-chu-min*) to preserve their culture. He gives the reader an indepth perspective of issues of culture and identity by focusing on Taiwan's earliest settlers—now a beleaguered minority—and shows us how they are trying to hold on to their culture even as it is under assault by the processes of mandarinization and globalization.

In chapter 6, Paul R. Katz explores the relationship between the scholarly debate over identity and the study of religion in Taiwan as it has taken shape over the course of the last five decades. He first introduces the larger issues and then provides a case study showing how these debates and arguments play themselves out even as scholars attempt to understand the nature of specific gods and the patterns of worship of these gods in Taiwanese cults.

Rubinstein's chapter (chapter 7) deals with the Ma-tsu cult in Taiwan and in Taiwanese temples. It examines issues of medium and message as they related to ways the leaders of three major Ma-tsu temples present their sites to the larger community of believers. He focuses upon the way these temple committees use time—temple history, space—the temple site itself, and words—the documents the temple produces and distributes—to make their own temple known. Their objective is to draw believers to them as they compete among themselves and with the other major temples that are devoted to Ma-tsu. An underlying theme in the essay is the way a temple's very structure and the actual presentation of a major temple's symbolic and material artifacts helps to foster the growth of the individual, community, regional, and ethnic/provincial identities of core temple members and of the pilgrims who participate in a major temple's annual ritual celebrations.

With Avron A. Boretz's chapter (chapter 8) we shift from the Ma-tsu cult and local cults to another form of popular religion, a martial arts troop participating in various community ceremonies. Boretz, both a scholar of *and* a practitioner of Taiwanese martial arts, focuses on a martial arts group in the area near and in his adopted city in the ROC (Taitung) and studies the links between identity and one form of performance-oriented popular religion. His objective is to have us understand the process of identity formation by examining individuals whose very identities and social roles are tied to participation in certain forms of martial arts related ritual activity.

P. Steven Sangren's essay serves to bring together many of the subthemes of this book. His contribution is a long meditation on a number of related issues centering religion, local consciousness, and identity. It is an essay that is buttressed, as is always the case in his essays, with solid fieldwork-based evidence, and it also demonstrates a deep understanding of the different terrain—geographical, spiritual, and intellectual—that he concerns himself with.

The subject of this book, identity/role, and its relationship to Taiwan's society, political environment, culture, and religious universe, is a complex one that has been the focus of a growing body of literature. It is our hope that by devoting this book to certain sets of linkages we can shed additional light on the core issue of identity and the larger issues of the nature of Taiwan's complex—and everchanging—societal and cultural/spiritual matrix.

Notes

1. The summary of events presented below is based on our own fieldwork experiences in Taiwan, as well as articles published in the local mass media.
2. For more on incense power, see P. Steven Sangren, *History and Magical Power in a Chinese Community* (Stanford: Stanford University Press, 1987). For more on the Chen-lan Kung and its pilgrimage, see Huang Mei-ying, *T'ai-wan Ma-tsu hsin-yang te hsiang-huo yü yi-shih* (*Incense and Ritual in Taiwan's Ma-tsu Cults*) (Taipei: Tzu-li wan-pao, 1994), as well as the chapter by Murray Rubinstein in this volume.
3. See the February 18 issue of the *Chung-kuo shih-pao (China Times)*.
4. In recent years role theory has become a basic element in the sociological school (and psychosociology school) of social interactionism. One recent book in this field is John P. Hewitt, *Self and Society: A Symbolic Interactionist Social Psychology* (Allyn and Bacon, 1997).
5. The list of materials is extensive. Among those I obtained in a brief search were "Functionalism, Identity Theories, the Union Theory," in T. Subzuka and R. Warner, eds., *The Mind Body Problem: The Current State of the Debate* (Oxford: Backwells, 1994); Rosenthal, chapter V. "Behaviorism, Physicalism and the Identity Thesis," http://csaclab-www.uchicago.edu/philosophy Project/sellars/rosent5a.html.
6. "All the world's a stage," Shakespeare's Duke tells us, "and all the men and women merely players. They have their exits and entrances and one man in his time plays many parts." The Duke lays out the series of parts in seven stages that an individual plays on this great stage of life, from the infant to the adolescent to the lover—the young adult—to the soldier, "seeking the bubble's reputation even in the canon's mouth", then the mature man in mid career, "in Fair round belly, with good capon lined, with eyes severe and beard of formal cut, the old man—the lean and slippered pantaloon, with spectacles on nose and his big manly voice turning again toward childish

treble," to a final image of a man who is much like the child he was decades before. William Shakespeare, *As You Like It*, act II, scene vii

7. Studies on role and identity include: Blake E. Ashworth, "All in a Days Work: Boundaries and Micro Role Transitions" in *Academy of Management Review*, July, 2000; Shawn Megan Burn, Roger Aboud, and Carey Moyles, "The Relationship Between Gender Social Identity and Support for Feminism" in *Sex Roles: A Journal of Research*, June, 2000; Dennis A. Gioia, "Organizational Identity, Image, and Adaptive Instability," in *Academy of Management Review* (January, 2000).
8. This is taken from http//www.3.wcedu/-GUFSON/HBSE/THEORY/tsld016.htm. See also Catherine R. Cooper, "Theories Linking Culture and Psychology: Universal and Community Specific Processes" in *Annual Review of Psychology*, 1998.
9. Clare Cassidy, "Identity in Northern Ireland: A multidimensional approach" in *Journal of Social Issues*, Winter, 1998. Kelly H. Chong, "What it Means to be Christian: The Role of Religion in the Construction of Ethnic Identity and Boundary Among Second-generation Korean Americans" in *Sociology of Religion*, Fall, 1998; Ujvala Rajadhyayaksha, "Life Role Salience: A Study of Dual Career Couples in the Indian Context" in *Human Relations*, April, 2000, Stephen W. Floyd, "Strategizing Througout the Organization: Managing Role Conflict in Strategic Renewal" in *Academy of Management Review*, January, 2000.
10. The most useful overview of Confucianism and its impact upon China and East and Southest Asia is John H. Berthong, *Transformations of the Confucian Way* (Boulder: Westview Press, 1998). On Confucius and his thought see David L. Hall and Roger T. Ames, *Thinking Through Confucius* (Albany: State University of New York Press, 1987). For an analysis of Mencius see Kwong-loi Shun, *Mencius and Early Chinese Thought* (Stanford: Stanford University Press, 1997). On Confucianism and society in early Second Millennium A.D. China see Robert P. Hymes and Conrad Schirokauer, *Ordering the World: Approaches to State and Society in Sung Dynasty China* (Berkeley: University of California Press, 1993). On Confucianism and issues of role and identity see Pei-yi Wu, *The Confucian's Progress: Autobiographical Writings in Traditional China* (Princeton: Princeton University Press, 1990).
11. Steven R. Bokencamp, "Death and Ascent in Ling-pao Taoism" in *Taoist Resoures*, vol. 1, 2 1989, 1–21: and "Sources of the Ling-pao Scriptures" in M. Stricmann, ed., *Tantric and Taoist Studies* (Brussels: Institute belgede hautes etudes chinoies, 1983), 2: 434–486.
12. These chapters from the Chuang Tz can be found in Theodore DeBary, ed., *Sources of the Chinese Tradition* 1st ed. (New York: Columbia University Press, 1960). Scholarship on Taoism has expanded dramatically in recent decades. Two general works are Kristofer Schipper, *The Taoist Body* (Berkeley: The University of California Press, 1993) and Isabelle Robinet, *Taoism: Growth of a Religion* (Stanford: Stanford University Press, 1997). Works relevant to my discussion include Roger T. Ames, ed., *Wandering at Ease in the Zuangzi* (Albany: State University of New York Press, 1998),

Livia Kohn, ed., *The Taoist Experience* (Albany: State University of New York Press, 1993); Livia Kohn and Michael LaFargue, eds., *Lao-Tzu and the Tao-te ching* (Albany: State University of New York Press, 1998).

13. Those who are Buddhists in the United States deal with this contradiction. See the set of articles on Buddhism and American politics in *Tricycle* (Summer, 2000). On the relationship between Buddhism and the larger Chinese world see Jaques Gernet, *Buddhism in Chinese Society* (New York: Columbia University Press, 1995). On Buddhism and society in the Ming see Timothy Brook, *Praying for Power: Buddhism and the Formation of Gentry Society in Late-Ming China* (Cambridge, Mass.: Harvard University Press, 1993).
14. On the evolution of religious Taoism see Livia Kohn, *Taoist Mystical Philosophy* (Albany: State University of New York Press, 1991); Isabelle Robinet, *Taoist Meditation* (Albany: State University of New York Press, 1993). On ritual and the roles of ritual specialists in Taoism see John Lagerwey, *Taoist Ritual in Chinese Society and History* (New York: Macmillan, 1987). See also Michael Saso's popular and very entertaining study, *Blue Dragon/White Tiger: Taoist Rites of Passage* (Honolulu: University of Hawaii Press, 1990).
15. A key work for understanding the development of popular religion in China is Valerie Hansen, *Changing Gods in Medieval China* (Princeton: Princeton University Press, 1991). See also Kenneth Dean, *Taoist Ritual and Popular Cults in Southeast China* (Princeton: Princeton University Press, 1993), and Donald S. Lopez, Jr., ed., *Religions of China in Practice* (Princeton: Princeton University Press, 1996).
16. My own knowledge of role specialization and temple organization is derived from my reading of temple gazetteers, reading the new popular religious compendium published in Taiwan over the course of the last decade making observations of temple staff and specialists during festivals, busy weekends, and on ordinary work days, and conducting interviews with those who staff and maintain temples in the cities and the villages of Taiwan, in cities in Fukien and on Meichou and in the temples of Kowloon and the New Territories.
17. Terry Kleeman, *A God's Own Tale* (Albany: State University of New York Press, 1991).
18. The basic book on Chinese pilgrimage is Susan Naquin and Chu-fang Yu, eds., *Pilgrims and Sacred Sites in China* (Berkeley: University of California Press, 1992). On pilgrimage in Taiwan see Chang Hsun, "Incense Offering and Obtaining Magical Power," Dissertation for the Ph.D., University of California at Berkeley, 1993. See also Murray A. Rubinstein, "The Revival of the Mazu Cult and of Taiwanese Pilgrimage to Fujian" in *Harvard Studies on Taiwan: Papers of the Taiwan Studies Workshop*, vol. 1 (Cambridge: Fairbank Center for East Asian research at Harvard University), 89–125.
19. On the sectarian tradition on Taiwan see David K. Jordan and Daniel L. Overmeyer, *The Flying Phoenix: Aspects of Sectarianism in Taiwan* (Princeton: Princeton University Press, 1986). See also Ian A. Skoggard, *The Indigenous Dynamic in Taiwan's Postwar Development: The Religious and*

Historical Roots of Entrepreneurship (Armonk, N.Y.: M. E. Sharpe, 1996), chap. 8.

20. On issues of Taiwanese identity see Alan M. Wachman, *Taiwan: National Identity and Democratization* (Armonk, N.Y.: M. E. Sharpe, 1994). The impact of identity as an element in Taiwan's history is examined as a theme in the chapters of Murray A. Rubinstein, ed., *Taiwan, A New History* (Armonk, N.Y.: M.E. Sharpe, 1999). Identity politics also serves as a major theme in Hung-mao T'ien, ed., *Taiwan's Electoral Politics and Democratic Transition: Riding the Third Wave* (Armonk, N.Y.: M. E. Sharpe, 1996).
21. Questions of the evolution of identities and subidentities over the course of Taiwan's long history are examined in Murray A. Rubinstein, "Taiwan as China's Contested Maritime Frontier: Han Socio-cultural Expansionism and the Problem of National, Provincial, and Local Identities" (Presentation, Northwest China Council, Portland Oregon, March 1, 2001). See also the relevant articles in Rubinstein, ed., *The Other Taiwan*, Section I. That the point I make is applicable to the modern Taiwanese scene was demonstrated very clearly at a panel on issues of identity organized by the Taiwan Group of East Asian Institute at Columbia University held on April 23, 2001. After Rubinstein presented an overview of the evolution of levels of identity in Taiwan's history, students from Taiwan representing the three major ethnic groups on the island, the Taiwanese (*Hokkien*-speaking and the *Hakka*) and the *kuo-yu*-speaking Mainlanders, spoke not in scholarly but in very personal terms. Each discussed how they saw each other and how their own ethnic/subethnic identities affected their view of themselves and the larger Taiwanese world they grew up in. The very slippery and shifting nature of identity as social and self-actuating construct were on display during the actual presentations and the question and answer period that followed.
22. Hans Mol, *Identity and the Sacred: A Sketch for a New Social-scientific Theory of Religion* (New York: Free Press, 1976). "Introduction" in Hans Mol, ed., *Identity and Religion: International Cross-cultural Approaches*, Sage Studies in International Sociology, 16 (London: Sage Publications, Ltd., 1978), 1–17.
23. Abdullah, Taufik, "Identity Maintenance and Identity Crisis in Minangkabau" in Mol, ed., *Identity and Religion: International Cross-cultural Approaches*, 151–167. Lewins, Frank, "Religion and Ethnic Identity" in Mol, ed., *Identity and Religion: International Cross-cultural Approaches*, 19–38. Mol, *Identity and the Sacred*, 3–4, 13, 182, 183. Mol, *Identity and Religion: International Cross-cultural Approaches*, 2.
24. See for example the works of Timothy Brook, *Praying for Power: Buddhism and the Formation of Gentry Society in Ming China*; Yu Chun-fang, *The Renewal of Buddhism in China: Chu-hung and the Late-Ming Synthesis* (New York: Columbia University Press, 1981); Kenneth Dean, *The Lord of the Three in One* (Princeton: Princeton University Press, 1998); Stephen E. Eskildsen, "The Beliefs and Practices of Early Ch'üan-chen Taoism," M.A. Thesis, University of British Columbia, 1989; Vincent Goossaert, "Le creation du taoïsm moderne. l'ordre Quanzhen," Ph.D. thesis, École Pratique de Hautes Études, 1997; Daniel Overmyer, *Folk Buddhist Religion* (Cambridge, Mass.: Harvard University Press, 1976); Barend ter

Haar, *The White Lotus Teachings in Chinese Religious History* (Leiden: E. J. Brill, 1992).

25. Frank Lewins, "Religion and Ethnic Identity," in Mol, *Identity and Religion*, 3–4, 174–175.
26. Mol, *Identity and Religion*, 11–13, 15.
27. Murray A. Rubinstein, ed., *Taiwan, A New History* (Armonk, N.Y.: M. E. Sharpe, 1999).
28. Wachman, *National Identity and Democratization*.

CHAPTER 2
ON THE ORIGINS AND TRANSFORMATION OF TAIWANESE NATIONAL IDENTITY

Mau-kuei Chang

This chapter intends to explain the many sources of and historical changes in Taiwanese identity (*Taiwanren rendong*). But first it is important to explain what the word identity means in this chapter. The term national identity is usually referred to as feelings, sentiments, and bonds that people feel for their own country, or nation. But this emphasis on sentiment could be misleading in our understanding of the problem of nationalism or nationalistic conflict as mainly a reflection of emotional or primordial conflicts—or conflicts caused by the human need for belonging. In fact, national identity arises, or emerges for reasons that are much broader than sentiment and the need for belonging; and it is always constituted with normative discourse, argued and supported with forceful moral–political claims. This is the reason for people of nationalistic thinking genuinely believing that they are the "righteous" people with a justifiable moral base. The question of nationalistic identity, therefore, becomes the question of the moral horizons of the group of individuals who are considered to be nationalistic.

National identity is also interpreted by some as the consequence of the indoctrination of nationalism, which is nothing but a strategically created ideology. They believe that nationalism is a tool employed by nationalistic elite for the purposes of industrialization, the advancement of class power, the hegemonic domination of the populace, or the suppression of the "others."[1] It is very true that nationalism does have instrumental values, which could even lead to atrocities in extreme. People in such situations often call upon national identity to coordinate their actions. Consequently, it is crucial for scholars to be sensitive to the conflation between nationalistic claims and its analysis.[2] Not only should scholars be against the stereotypes and the bias people might have when making nationalistic claims, but they should also be faithful to the interpretation of the past that the nationalists usually want to appropriate for their purposes.

However, seeing nationalism merely as a political doctrine or an engineered nation-building project does not help us to understand the compelling moral force that more than often obliges people to undertake this doctrine. Nor can this help us to understand the significant meanings of existence that many people attribute to their nationalistic standing. For our study of Taiwanese identity, I think that Charles Taylor's recent works on the formation of the identity and the making of self in the modern era are helpful. For instance, Taylor argued in his book titled *The Sources of the Self* that first, an individual's "self" can be understood only in terms of its "narratives," the stories that can be told about oneself. And the "self" can never be understood in its abstract meaning.[3] Second, the answer people have for the question "who am I" cannot escape from the moral domain of that person. The answer for "who am I," or "what is my identity" can never be understood as a mere description of a static "fact." The answer actually "provides the frame within which they can determine where they stand on questions of what is good, worthwhile, admirable, or of value."[4] For Taylor, self-identity is understood as a moral framework, a horizon of meanings through which people make out the significance of things that they have experienced, and things that have surrounded and constituted them. Taylor believes this is so because the fundamental quality of human agency cannot but "orient to the good." Only under conditions of humiliation or suppression, do we find people forced to compromise their integrity with self-denial and self-mutilation. To situate one's past experience in relation to "the good" thus provides meaning for one's life, and determines the direction of one's life as well.[5]

Taylor asks us to study the moral aspect of nationalism and national identity. We cannot limit our understanding of nationalism to just its instrumental value or to its ideological nature. A fuller understanding of national identity requires the recognition of this normative orientation. Only by recognizing it, can we understand the collectively compelling aspect of national identity. The research question of national identity in this chapter is similar. It is shifted away from both the study of primordial feelings or political ideologies, to the study of the ontology people have for themselves in relation to the national question under different historical conditions.

The second term we need to clarify is the term "nation." We all know that it is an elusive term with many meanings in various Western languages. Its Chinese equivalent *minzu* is the same. Chinese nationalists first borrowed the term *minzu* from the Japanese modernists during the late nineteenth century when the Social Darwinism was still prevalent. But the term has been attributed to with different meanings. In its early usage it contained the notion of people from the same lineage (or "race," *zhongzu*). And this meaning is still remembered occasionally, when used by ordinary people, especially among the older generation, in their political discussions. The term can also mean state-nation (*guozu*), the people as a whole consisting of many subnational groups in a multinational state, as many contemporary social scientists would prefer. It can also refer to a culturally or supposedly "racially" distinctive people, such as a minority or an indigenous people as it is used in mainland China now. On the other hand, it can mean the compatriots in general or "all" of the citizens. Or, it can refer to the political entity that governs the people (the sovereign country, or a synonym when referring to *the* government). As the historical context shifts, or the situation changes, the use of the same word can have different meanings. It is unfortunate that we will have to bear with these ambiguities from time to time as this discussion develops. This is so because discourses of what constitutes a nation (whether that is a Chinese or Taiwanese nation) have taken on so many forms and directions. People in Taiwan don't necessarily agree with each other about what the term means for them. And, surely, it can be used just for making different political claims. I don't think there is any way that we can get around this ambiguity when exploring the origins and development of Taiwanese national identity.

In short, this chapter suggests that the study of Taiwanese identity should be about the study of the origins and changes of the meaning of existence, the normative discourse, and the "correct" directions or strategic actions people believe they should take to define themselves. When we ask what it means to be a nationalistic Taiwanese, we are actually asking what the moral horizons are for a nationalistic Taiwanese. As I have said earlier, since there is always an intention involved when people try to argue for a distinctive and categorical identity, we must also recognize and explore the divergences of the origins, such as who were the leaders, and what were their class and ethnic backgrounds. In this chapter I will argue that the origin of Taiwanese nationalistic ideas can be traced back to at least the 1920s. But the "maturation" of Taiwanese identity, or the convergence of many sources, did not take place until the late 1980s, and it is still not a fully grown nationalism in the present day.

I hope I can show also that this convergence and development is not an inevitable consequence of ethno-conflicts nor an asymmetrical ethnic relation between the mainlanders and Taiwanese, although it is now generally interpreted as such.

Being Han-Chinese Settlers in Taiwan

It is important to note in the beginning that before any of the settlers, Taiwan was inhabited by different aboriginal peoples.[6] They are the "real" natives who have lived in Taiwan beginning from approximately 6,000 years ago. According to Taiwan's civic registration, still there are about 340,000 indigenous people in Taiwan, equivalent to 1.6 percent of the total population. It is unfortunate that the people who are normally addressed as the "native Taiwanese" by foreign press and political scientists does not refer to the indigenous population but to the descendants of early Han-Chinese settlers in Taiwan. The early settlers migrated to Taiwan beginning from the late sixteenth century from Hokkien and Canton Provinces in southeast China. In 1684, Emperor Kangxi of the Manchu Dynasty made portions of central and southern Taiwan a prefecture under the jurisdiction of the Hokkien Province.

Two studies about the Han settlers in late eighteenth- and nineteenth-century Taiwan are important for our understanding of how people were organized and what their identities were. One study was done by Li Kuo-ch'i about the "*neidihua*" of Taiwan around the 1880s.[7] *Neidihua* refers to the process by which Taiwan was elevated by the Manchu government from its low priority and low-level ranking in the Empire to become like the more "civilized" or established part (the "in-land") of the mainland, or to become like other "normal" provinces in the mainland. At that time, able Manchu officials were sent to Taiwan to develop infrastructure and industry, explore natural resources, acquire aboriginal territory, and promote the "pacification" and "culturalization" of the aboriginal populations. The political goal of the government was to build a stronger national defense of this island to fence off foreigners' encroachment along the southern and eastern coast of China.

The other study was done by Chen Chi-nan about "*bentuhua*" or the localization and indigenization of the Han settlers.[8] Citing evidence of changing patterns of intergroup rivalries or conflicts (*xiehdou*), he was able to demonstrate that after settling in Taiwan for more than one century, people gradually developed new territorial identities rooted in Taiwan. They were now less divided by their differences in ancestral lineage linked to different hometowns in the mainland, than by their pragmatic interests around their territorial and local bases, whether that was a village, alliances of several villages, or a regional township. In short, the settlers began to identify with their settlements in Taiwan and make their settlements their new homeland.

What can we derive from these two studies in relation to the later development of Taiwanese identity? The answer is, without a doubt, that the Chinese and Han civilization was the ruling culture that prevailed on top of the settlers' society. As Li guochi argues, Taiwan was becoming more "civilized," abandoning its "aboriginality" and frontier status, so to speak, from the viewpoint of high culture and the political authorities. Chen Chi-nan depiction that social attachment to regional community in Taiwan was already developed among the settlers was also true. Li looked at the changing politics from above, while Chen looked at the changing of social organization and the formation of local identities from below.

Today, some Taiwanese nationalistic historians and politicians claim that the concept of "being Taiwanese" or "defining one's self as Taiwanese" is about "four-hundred-years" old (since the late sixteenth century). This cannot be true since the collective imagination of "being Taiwanese" cannot be found before 1895. Before 1895, both the first settlers and their descendents were all looking at China as their place of origin, as the political center, and as their source of cultural practices and ancestral lineage. They were, rather, the *Hanren*, descendants of the Han, who practiced Han ancestor worship and other folk (or "popular," *minjian*) religious practices. Social communication was very limited due to the lack of modern transportation systems. The great majority of the settlers were confined to an agrarian economy that depended on the fertility of the land and wide-spread use of irrigation systems. Their identities were, therefore, very regional, bonded to the here-and-now, to their land and their villages in Taiwan. As for the better-off families, they were eager for imperial titles and official recognition. They pushed their children to study for the official examination of Han classical (*guwen*) literature and Chinese/Confucian moral classics (the *Sishu*, or *Four Books*), in order to gain power and high status. They used generous donations as means to obtain titles. In conclusion, the settlers as a whole could not perceive themselves as one larger "imagined community," or one larger group called "Taiwanese" that could be conceived of as substantially different from the Han-Chinese.

By any measure, Taiwanese nationalistic identity was not forged during that time, nor could it be called as such. The historical and socio-cultural conditions were just not there. It is arguably true that even in mainland China there was no significant Chinese nationalistic project until the early twentieth century. There were, in place, many ideas and discourses about saving China from foreign encroachment, and the ways to build China into a "wealthy" and "strong" nation since the late nineteenth century. But the Chinese intellectuals, in general, were at the most only in a process of defining themselves as a "coherent national intelligentsia," and not yet awakened to the Chinese Nationalistic call before 1895, the year of defeat and humiliation at the loss of war with Japan.[9]

The Awakening to Alien Power and Identity Crisis

China and Japan went to war in 1895 because of disputes over a political crisis in Korea. China was defeated convincingly by the new rising power in Asia. Taiwan was ceded to Japan in the *Treaty of Shimonoseki* as Japan's first major trophy on its way to imperialism.

The Han agricultural settlers, who were tied to their land and their territories, had responded with bloody resistance to the invasion of Japanese troops in the first two years. They were referred to as "the righteous army" (*yichun*) consisting of mainly ordinary people, led and supported by gentry and landlords from the same region, or from some regional alliances.[10] And yet their resistance was anything but nationalistic. As one famous Taiwanese writer, the late Wu Chuo-liu, has suggested, these desperate, nearly hopeless, and yet very brave people were not guided by modern national consciousness, but rather, by traditional Han folk consciousness when defending their own hometowns. They had no escape but to do the "right" thing, to go to war against foreign invaders. Their defensive actions were not different from their ancestors in the past when confronted by intruding bandits or rebels.[11] And they were not armed with demands for dignity, respect, or self-determination of the type that Taylor had theorized would fit a situation of this type. Their resistance actions "become" nationalistic only after such actions were interpreted as such by present-day nationalists.

The frequency and scale of armed resistance dwindled in the following years but continued to take place until 1915. After the gentry and peasant rebellion had been subdued, a more "marginal sector" of the society, such as people lacking property, continued to maintain the anti-imperialist struggle when pressed by the Japanese police. They originated from groups with strong solidarity and, sometimes, of devotees to one or another of the Han popular religious sects. In general, however, this kind of resistance was non-nationalistic except for one special case, when its leader was actually inspired by the Republican Revolution of 1911 and the actions of Kuomintang in southern China.[12]

The Japanese looked down upon the backwardness and the lack of a modern outlook of the Chinese settlers who had been abandoned by the Qing government. Within the new Japanese Empire, Chinese in Taiwan was regarded as a different "race." Japanese had employed brutal measures to crack down resistance, and had built a system that was protected and sponsored by the colonial government for the exploitation of both the Taiwanese farmers and the aboriginal tribal peoples in the interests of Japanese conglomerates. Armed with the special provision of the notorious "Bill no. 63" (1896–1920), and the "3-1 Bill" which followed, the Governor General of Taiwan (*Taiwan Sotoku*) was largely free from any of the accountability to the Imperial Diet in Tokyo. The Governor General, the supreme commander of the military and police, was the highest administrator, and the sole lawmaker for the new colony and thus was, virtually, the "emperor" of Taiwan. Ordinary Taiwanese people were organized according to the civic registration system—the *baojia* system—and under that system (one that had been adopted from the Chinese system of the Qing), neighbors were held accountable for each other's unlawful activities. Death sentences were given, except for a few exceptions, to anyone who was found to be involved with armed resistance under the terms of the infamous *Decree against the Bandits.*

But from 1914 to 1918, progressive minded Japanese political activist *Itagaki Taisuke* spoke for the assimilation of the Taiwanese "race" into the Japanese "race," and for the eventual "normalization" of Taiwan's status as a "natural extension" of the more established part of Japan. It was argued that both the Taiwanese and Japanese shared many "racial" similarities. Once the Taiwanese would have achieved "an appropriate civilization level like the Japanese" (through some nationalistic educational program, perhaps), and learned the civic virtues and loyalty to the emperor, Taiwan and the Taiwanese could eventually become no different from, or become equal to the rest of Japan and the Japanese.

It was attractive for the colonized Taiwanese but only until the early 1920s. About five hundred people (mainly Taiwanese but also some Japanese) joined together to organize the first modern Taiwanese political organization—*Taiwan Dokakai* (or *Taiwan Tonghuahui*, the Taiwan Association for Assimilation)—to promote these ideas. These Taiwanese had hoped to work with those Japanese who were sympathetic to Taiwanese desires to achieve equality under the new circumstances. Of course, the assimilation argument the Japanese presented could not possibly offer the Taiwanese real respect. It was, rather, the equivalent of asking the Taiwanese to admit their deficiencies and to accept their backward status. Like assimilation strategies everywhere, the Japanese policy of assimilation was inevitably hypocritical: It pre-supposed the existence of a long-term hierarchical relationship under the pretense of achieving eventual equality. The desire for assimilation actually reflected the humiliation of the first-generation Taiwanese elites under the Japanese. It was the consequence of their self-denial and fear of condemnation from their conqueror. Segregated from their cultural and political roots—China, their traditional Han civilization, as it existed on Taiwan, and their larger worldview—their sense of being "Chinese"—were smashed and could not be rejuvenated in the face of such a total defeat. Though ordinary people still maintained Han traditions and morality for their basic group identification, this traditional knowledge and social sensibility now could not help the elite classes to re-orient themselves to the unprecedented colonial situation that had been created by such a powerful intruder.

Contemporaries tended to interpret the Taiwanese elite's shortlived fantasy for assimilation into Japan, and for the desire to become equal citizens of the Japanese Empire as these elites' first strategy of resistance. But it can be better understood as a sign of a collective identity crisis among the Taiwanese elite under the Japanese. These elites that were trying to get hold of themselves—to recover their dignity and sense of honor in the face of a massive defeat that was so humiliating that it placed pressure upon them to deny their own past.

Different Paths to Home Rule

If we look at the decade from 1918 to 1928, we can find that from the youthful to the urban middle-class, from the great metropoles of Tokyo, Shanghai, and Peking to cities in Taiwan like Taichung and Taipei, Taiwanese elite were eager to look for new means for self-organization, and to a new language to help them understand and deal with their past humiliations and their existing set of depressing every day realities. This decade thus became a time of new ideas and new political thinking not only for the Han-Chinese in Taiwan, but also for those in China and in the heart of the colonial empire that was Japan.

Without any doubt the Taiwanese still considered themselves as basically Hanren, but the Han-Chinese tradition they used to cherish was no longer useful to them within the new colonial context that they found themselves in. And as they were all too well aware, on the other side of Taiwan Straits, the Chinese people had begun transforming themselves in the fluid political and social conditions of a China under the successive regimes created by the 1911 Republican Revolution. China was now taking on a historical trajectory that was vastly different from Taiwan's existence as Japanese colony and a part of an evolving Japanese sociocultural order: It was struggling to become a "modern" nation-state. Taiwan's situation was seen as impossible to reverse by both Chinese and Taiwanese at that moment in time: Thus the Taiwanese people, the new elites believed, had no other choice but to accept the status of being a Japanese nationality, and being a different "racial" group, but one of Chinese origin under, the Japanese Empire.

Although from many different places in Taiwan and in Japan, and with diverse ideological and class backgrounds, Taiwanese activists involved in resistance to the Japanese shared both common sets of knowledge and commonly shared sets of beliefs. These can be summed up as follows: (1) for the first time there was the general understanding of the fact that Taiwan was a colony and of the harsh reality that the Taiwanese people were being oppressed as a *colonized people*, or nation; (2) it was also understood that the Taiwanese people needed to struggle for "home-rule," whether it meant the creation of a representative council (as suggested by the right-wing reformists), or the total independence and liberation of the Taiwanese Minzu (as suggested by the communists); and (3) in order to accomplish "home-rule," these movements needed the involvement of ordinary people (or the masses). It should be clear that both the origins of the idea of Taiwanese nationalism and the ways of turning that idea into substantive reality were conceived during this period. This process can thus be seen as a search for a defined meaning of existence by many Taiwanese activists who were attempting to re-orient the "self" to the common good of a group of people—or polity—now called the "Taiwanese nation." Their efforts coincided with Taiwan's colonial conditions and anticolonial ideas developing in this region, as well as in the larger global context of the 1920s, which I will discuss in the next section.

Resisting Colonialism and Taiwanese Identity

Many Taiwanese activists of the 1920s and 1930s were from a younger generation of Taiwanese intellectuals, who were fortunate enough to go to the metropolis to learn about modern ideas and Western thinking through Japan's mediation. In the year 1920, there were about 400 young Taiwanese students in Japan, and the number increased very quickly to 2,000.[13] Tokyo became a major source of ideas for the Taiwanese home-rule movement. It is important to recognize the significance of the expansion of the horizons of young Taiwanese intellectuals who were not just looking back to Chinese traditionalism like the older generation.

In the metropolis, they were able to discover that the oppression that Taiwanese suffered as a result of colonialism was not an isolated phenomenon. Rather, they discovered that it was a part of the situation common to that experienced by those in many of the colonized nations that were found in the world of the 1920s throughout the world. And the solution to this problem was found in the most prominent philosophy for ending colonialism after World War I—the Wilsonian principle of self-determination. Further, it was thought that the League of Nations (nation-states) would become the international forum for the maintenance of the peace in the new world. In fact, it was an era of the dissolution of many of the old empires in Europe. By coincidence, it was also a democratic era in Japan, the era of *Taisho Minshu* (democracy in the Taisho years). Japan was trying to transform itself from a centralized bureaucratic state, forged in the Meiji period, into a system of parliamentary party politics and democracy. Progressive and democratic Japanese figures had also entertained the idea of equal rule or self-governance for their colonies when questioning the government's colonial policy.

It is clear that since the 1920s, people had already begun to define the Taiwanese people as a collectivity, a distinct "nation" ruled by the Japanese colonial government. "Taiwan," as a signifier for the place and the people as a whole, had been widely adopted. It appeared in the names of many social and political organizations, study groups, and in the political analyses of Taiwan's colonial conditions. The term "Taiwanese" appeared in their writings, their arguments, their feelings for, and in their plans for collective actions. The image of the Taiwanese as a whole, a people with distinctive national qualities, was surely conceived in their minds at this point in time.

But when arguing for the distinctiveness of the Taiwanese, these activists were also very clear and quick to resort to Han-Chinese tradition, their basic roots. The term "nation" in their minds was like a cultural-racial category providing the "natural" and the "real" meanings of existence under the Japanese. The term was much less like the "created" or "man-made" political notion calling for Taiwan independence, which took place later in the post–World War II period. And this call for distinctiveness was definitely not relevant to the calling for an unequivocal separation from China and the Chinese.[14] As lifelong Taiwanese nationalist and sociologist, Huang Zhaotang, wrote about the resistance groups back then: "even if they promoted the idea of Taiwanese independence, and their notion of Taiwanese Nation was clearly distinct from the Japanese Nation, the Taiwanese Nation was not to be different from, and were mixed up with the Chinese Nation."[15]

The call for the collective existence of the Taiwanese at that time could be understood not just as an inspiration for a distinctive national identity, but could also be seen in putting them in an ethno-national "sandwich." People were stuck between their Chinese heritage on the one hand, and their institutional Japanese nationality on the other. They could not feel comfortable being Chinese, because they were cutoff from China, and were despised for having Chinese roots. On the other hand, they could not be Japanese either, for they were treated as unqualified colonial subjects pressed into assimilation.

The consciousness of this dilemma was articulated succinctly by renowned resistance activist and medical doctor Jiang Weishui during the founding of the first Taiwanese movement organization, the *Taiwan Cultural Association* (*Taiwan Wenhu xiehui*) in 1921, itself an epoch-making gathering of Taiwanese intellectuals and middle-class professionals. Jiang and his peers felt that the condition of the people in Taiwan originated from the defeat of China in 1895, and the continuing hostility between the Chinese and Japanese. The Taiwanese were a branch of the Chinese nation forced to take on Japanese nationality with lower status in Taiwan because of the conflict between the two large nations. Since then the Taiwanese were trapped in the middle and were thus victimized; If Japan and China could become friendly toward each other, not only the Taiwanese, but also Asia and the entire world would have benefited.[16]

One of the most noted and lasting home-rule movement was the Petition Movement for Taiwan Council (*yihui qingyuan yundong*). It lasted from 1921 till 1935. It was begun by groups of Taiwanese youth in Tokyo. The movement was then pushed forward by the Taiwanese gentry and by those members of the new middle class who had received a Japanese education and who then had become white-collar professionals—bankers, doctors, teachers, and writers—in both Taiwan's urban zones and in the rural townships. Constrained by their class origins, they often distanced themselves from the more radical socio-political movements of the left-wing Taiwanese. Famous leaders among them were Lin Xsiatang, Yang zhaojia, and Jiang Weishui, all individuals who first participated in this movement before moving into the leftist camp. They had engaged in activities like opposing the exclusive and the discriminative "Bill no. 63," lobbying and petitioning the Imperial Diet and the Governor General for the establishment of an elected council representing the Taiwanese. Such a council would have served the government and allowed the Taiwanese leaders to speak for the Taiwanese majority and balance *Taiwan Sotoku*'s supreme authority. They had also published newspapers, magazines, and organized lectures and drama tours around the island to promote the idea and the necessity of a Taiwan Council to the public. In general, their resistance strategies had been modest and elitist and their activities had been undertaken with sympathetic support from more progressive Japanese intellectuals and Diet members.

This movement fell apart soon after the escalation of the second Sino-Japan War in 1937. Pressure from rising fascism in Japan and the *Kominka* movement (cf. later) in Taiwan was too heavy for them to carry on any social or political reform or a form of negotiation with their Japanese masters. The colonial government managed to appoint leaders of the movement to symbolic council positions and tried to win their endorsement for Taiwanese assimilation with honorable status. But the Japanese never granted their wishes for a genuine Taiwan Council. To entice Taiwanese assimilation, the Japanese authorities were willing to offer Taiwanese elites only limited local political participation through various appointments and through indirect election in the last two years of its rule in Taiwan.

We need to take note that Japan was not the only place where the search of Taiwanese national identity was taking place. Taiwanese people were also inspired by the Republican Revolution in China since it had begun in 1911 and the movements that evolved out of that event such as Sun Yat-sun's development and promotion of an ideology of a quasi-Western-style Chinese nationalism, and the *May Fourth Movement*, with its stress upon a new Chinese culture oriented toward Western values and structures and its promotion of a more sophisticated, elite-led nationalism. The spread of Stalinism and Communist internationalism after World War I in Shanghai and Tokyo also influenced Taiwanese leftists like Xie Xuehong. These impacts on the entire Taiwanese resistance camp, now often played down by Taiwanese nationalists, were very significant.

This was especially true for the Taiwanese Communist Party established in Shanghai in 1927. The party had adopted an internationalist platform—a united front of all oppressed and colonized nations to overthrow feudalism and imperialism. Its ultimate goal was to strive for Taiwanese national independence (*Taiwan minzu duli*) and overthrow the Japanese. It was actually the first political program that advocated an unequivocal Taiwanese nation and independence. As Chen Fangming[17] has written convincingly, the Taiwanese Communist Party and their secret network were very active in penetrating many of the sociocultural movements in Taiwan, ranging from the farmers' association, workers' strikes, and other cultural and political resistance groups.[13] It should not be a surprise to us that the early Taiwanese leftists and the modest opposition who opted for the system reform with a Taiwan Council were often at fierce competition and at odds with each other.[19] But the conservative wing appeared to have more visibility and enjoyed wider claims at that time because their petition activities were often within the tolerance limit prescribed by the colonial government.

Reaching Out to the Populace: A Language Hurdle

For those who have had the privilege of access to metropolitan ideas, plus some others in the newly emerging middle strata of society, the notion of a Taiwanese people struggling to obtain control of its own destiny has gradually become significant. They acquired the necessary level of linguistic sophistication to promote strategies redefining their own existence under the colonial situation, and the ability to work on their identity problems in relation to China and Japan But the general public found it much more difficult to comprehend. The worldviews or the sociocultural perspectives of the majority of Taiwanese before 1915, were not very different from those of people in China. More than 80 percent of the adult population were poorly educated farmers. They were living in poverty with little opportunity for education. They were also exploited by both Japanese colonial economic interests and large Taiwanese landlords. The greatest obstacle to reach these illiterate masses was, of course, the suppression of political activities in Taiwan by the Japanese police. The other problem was the language gap between the elite and the masses.

Very few studies have asked this question: How did or could the more enlightened class spread their interpretation of the meaning of being Taiwanese and their ideas about home rule to the majority of Taiwanese, who ranged from rural gentry and small landlords, to traditional Han literati, to the illiterate masses of farm workers and peasants? Only one newspaper, published by activists in Taipei, and with the largest daily circulation being around 2,000 copies, could be thought of as a major forum for public discourses. And, after 1937, the Taiwanese were completely forbidden to use Chinese characters. Before 1937, the major hurdle for the diffusion of ideas was the problem of finding an adequate writing system that could be easily comprehended by the less educated.

Except for the aboriginal groups, the two common dialects used most in Taiwan were the southern Hokkien (or the *Hoklo* dialect) and the Hakka. These two dialects (with many accent variations) were mutually unintelligible and were without written form. And the traditional writing system used in Taiwan had been classical Chinese, which was accessible only to the literati minority and not comprehensible to the masses. Encouraged by the *May Fourth Movement* in China, young repatriating Taiwanese writers were eager to follow suit. They started a shortlived yet blossoming vernacular movement (*xin baihuawen yundong*) in Taiwan in the mid-1920s, hoping to revolutionize the way of writing and change feudalistic thinking in Taiwan. They did succeed in some aspects such as forcing the classical form of Chinese to a retreat and introducing more cosmopolitan ideas and writing styles to the younger generation writers. But still they could not succeed in reaching out to the populace directly. This was because the new vernacular writing system was based on Mandarin Chinese, originated from northern China. It was neither based on the southern Hokkien nor on the Hakka.

It was a time when a small number of Taiwanese intellectuals began to experiment with Romanization systems for writing down the Taiwanese dialect (*Taiyu*, referring to the southern Hokkien). A Romanization system developed by the English Presbyterians was being used by the members of the Presbyterian Church on Taiwan but this church was relatively small and had limited influence at that time. Thus the larger Romanization effort did not succeed. The gentry, the literati, and the new class might have been content with the use of Chinese and Japanese characters which the youth began to acquire in the Japanese primary school from the 1920s. Some Taiwanese had called this younger generation as the "*Taishou* Men," or the young man who grew up in the years of *Taishou* Emperor. There was a vested institutional and class interest or bias in favor of the status quo. In any case, by the 1930s, Japanese had ironically became the *lingua franca* for educated people across all dialect groups in Taiwan.[20]

Writings about and for the masses—for the less privileged—was the main goal of a new wave of Taiwanese literature from 1930 to 1937 (*xin wenxue yundong*). Lai He and Yang Kui were two renowned Taiwanese writers widely commemorated even today by many. Lai He, credited as the founding father of Taiwan new literature, wrote in vernacular Mandarin in his early years and received much acclaim. But he was frustrated that the people he wrote about could not understand what was written about them. He stopped writing for a while and then began again, this time experimenting with varied combinations of the southern Hokkien vernacular. But then even less people understood. He then adopted the use of *Zhu-zhi-diao* (a folk rhythm easy for recital) for writing. Yang Qui was another example. He is remembered as the most stubborn of the anticolonial writers in the 1930s and 1940s. But all of his best works were written in Japanese without exception, and he was highly praised in the Japanese leftist literature circle before the escalation of the second Sino-Japan War. The language hurdle was never fully overcome, since the use of Chinese characters and dialects were outlawed as the *Kominka* movement intensified after 1937. Since then, all Chinese books and newspapers were banned from publication in Taiwan, as well as imports from China.

The activists and intellectuals had also organized many lecture or drama tours to promote their ideas and political agenda before the ban on Chinese. They were probably well received by the audience, but they were also under constant surveillance by Japanese police. It is not difficult for us to imagine the pressure on the speakers, performers, and the audience. It was impossible to inspire the general public to rally around an independent national identity for Taiwanese. In fact, it was never even discussed in public. The most often discussed theme was, rather, to awaken the Taiwanese to their own rights to gain a proper status of respectful citizenship (of Chinese–Taiwanese origin) in the Japanese Empire.

What is the importance of the language hurdle for this discussion? It helps us to understand that the newly emerged Taiwanese identity could not reach a very large audience. And there were few alternatives to Japanese education and Japanese language to "enlighten" the general population. It was ironic that the cultural awakening of the Taiwanese had to rely on the use of the colonial language, and not their mother tongues. This would also help us appreciate why the Japanese education was successful in nurturing Japanese patriotism and Japanese inclinations without any serious challenge from the Taiwanese general public in the later decade.

Factional Competition and Class Origins

Taiwanese activists were further divided along class origins and differences about what the Taiwanese needed to change and to coordinate for their collective good. This was marked by the split of the *Taiwan Cultural Association* in 1927, and the split of the *Taiwan Min-Chung Tang* (People's Party)[21] in 1930. Both splits were similar in nature. The leftists, inspired by communist ideas from Shanghai and Tokyo, or by the southern revolution in China, went for social, economic, and national reform at the same time. The communists had attempted to mobilize the peasants and the workers very discretely but not with large-scale success. Yet, the right-wing reformists, consisting of a well-educated middle class and gentry, inclined toward more conservative political programs for reform, such as petitioning for local rule from within the system. They competed intensely for the support from the Taiwanese. The left-wing openly criticized the right-wing for selling out the interests of the Taiwanese masses in exchange for personal gains and government recognition. According to Japanese police records, the right-wing lecture gatherings were often disrupted or broken up by the left-wing hectors.[22]

A comparison of this conflict with the development history of different roads of Chinese nationalism before 1937, reveals that Taiwan's situation was somewhat parallel. Simply put, in mainland China, there was bourgeois nationalism led by the Nationalist Party and socialist nationalism led by the Chinese Communist Party.[23] There had been periods for cooperation and overlapping of ideologies and personnel, but they inevitably ended in conflict and breakup. The intra-conflicts were about what were the just choices to make for class alliances and about what class was to lead China to confront imperialism. But the differences within the Taiwanese resistance were more. They were more to do with just choices to make, but were also about the question of national identity under the Japanese, or the question "to be or not to be." On the other hand, the Chinese had less identity crisis than the Taiwanese. The Chinese had few doubts about their meaning of existence as one large nation, the Chinese "race" (*zhong*), because their past and lineage had never been forcefully erased by powerful foreign rule. The Taiwanese, on the other hand, were less lucky. Their past was discriminated against by alien colonization; they had to deny and to compromise themselves to regain meager recognition from the dominant Japanese. They were in fact "sandwiched" people of some unknown destiny, which they could not even discuss freely in public.

On the other hand, the competition and division among Taiwanese activists from the late 1920s to the early 1930s demonstrated well that a common enemy alone was not enough to create a unified national image of Taiwan. Taiwanese identities were thus articulated through different paths of political actions and strategies, and were, therefore, given different meanings. It seemed that not any one of these factions from the resisting camps could win enough support to arrive at a consensual Taiwanese identity under the Japanese system. If they were not silenced by the Japanese in the 1930s, and if they were not suppressed by the right-wing Chinese Nationalistic government after 1949, Taiwanese nationalism might have gone either way: the bourgeoisie nationalistic road or the communist nationalistic road. Its future was by no means "destined" in the 1930s and 1940s. The discourse and narrative of the Taiwanese nation would have steered through many possible courses if not for the later political developments.

Becoming "Japanese-Taiwanese" or "Taiwanese-Japanese"?

In 1994, in an interview titled "The Sadness of Being Taiwanese" by the late Japanese writer Shiba Ryotaro, Taiwanese President Lee Denghui was quoted as saying "I thought that I was a Japanese until I was twenty-years-old." Lee believed that he was speaking for his generation as a whole. He was certainly right. The Japanese were very serious about promoting assimilation and patriotism when modern national education was institutionalized in Taiwan in the 1920s. In competing for Taiwanese' minds, the Taiwanese resistance did not seem to have had the upper hand.

Taiwan under Japan had achieved the highest industrialization level in Asia except for Japan by 1920. North–south roads, railways, telecommunications, electricity, radio, museums, unified monetary and measurement systems had helped to develop an image of the islanders and the Taiwanese as a larger existence. Common people were no longer confined to their local or regional identities. The development of tap water, modern medicine and professions, public schools, agricultural inventions, sciences, and so on had helped to improve the living standards of the Taiwanese. The scientific knowledge employed by the government for administration, the discipline of tough but able bureaucrats and police, the institution of law and order, the introduction of the new concept of duty to the emperor, and the elaborately designed political rituals for patriotism, and the like, had left deep impressions on the minds of the Taiwanese.

Two Taiwanese nationalist intellectuals exiled to Japan in the 1950s, also shared similar ambivalent feelings about their former colonial master. The late Dr. Ong Jok-tik iterated that the most abhorring part of Japanese education was the indoctrination of patriotism and the loyalty to the Emperor. But Japanese education also lifted Taiwan from a superstitious and feudalistic society to modernity. In his assessment, the Taiwanese became learned and modernized people because of Japanese education.[24] Dr. Huang Zhaotang, now a university professor in Japan, wrote that the Taiwanese had a bloody hatred for the Japanese, and he personally could never forget the shame that the Japanese had brought upon Taiwan through racial colonialism. But he was also quick to point out that the Taiwanese were much better off materially under Japan.[25] He also believes in the possibility that being an assimilated and educated Japanese-Taiwanese, and being a part of the great empire gave the Taiwanese a chance to feel proud of themselves by abandoning their old, low status of Chinese identity.[26]

Being pragmatic and cooperative would have been just passive compromises of individuals in the face of great adversaries. But it is another matter when some Taiwanese began to identify with the Japanese eagerly. Why was it so? The Japanese had managed to achieve a certain degree of success through its *Kominka* movement in 1935. The movement was prompted by the ambition of the Empire to advance to southeastern Asia. Taiwan was regarded as a springboard of this advancement. The assimilation and loyalty of the Taiwanese (more than 5 million at that time) were needed for this strategy to be successful. The movement included social, cultural, and linguistic policies to facilitate the rapid learning of Japanese, the sciences, and Japanese virtues. Chinese was forbidden in public occasions, and banned entirely in print. People were persuaded to abandon their traditional cultural religious practices as "un-scientific" or "un-patriotic," and they were also encouraged to change their Chinese surnames into Japanese and abandon their ancestor worship for Japanese Shintoism.[27] The government had set up "exemplary Japanese-speaking families" to honor the cooperatives with social recognition. The assimilating elites were also given honorable status, and, appointed to positions to lead the movement. Huang Zhaotang said that the streets were full of true believers that were proud to forget Taiyu and to speak the "national" language.[28]

After 1937, Japan began to prepare the Taiwanese to join the War. For the Taiwanese to join the Imperial Forces, risking one's life for Japan, became the ultimate test of their loyalty and assimilation. As for the Japanese, to allow the Taiwanese to fight side by side with them became the ultimate test but also for their trust in and equal treatment of the Taiwanese people. The hegemonic success of the *Kominka* movement had readied some Taiwanese to do just that but not the Japanese. Hesitant to trust the Taiwanese or offer them equal status, Japan did not institute an all-out draft system until the last year of the War, despite the rising pressure for new recruits from the battlefields. Taiwanese youth were admitted into the Imperial Forces on a voluntary basis only after 1942. And they were admitted by their degree of assimilation, level of civility, loyalty, and physical condition. Otherwise, they might be qualified only as auxiliary military workers or interpreters on a contractual basis. The Japanese government reported that about 17,000 young Taiwanese men joined the rank of voluntary soldier out of fierce competition. And about 120,000 men were hired as auxiliary laborers to work for the military.[29]

Taiwanese intelligentsia and modern artists were also encouraged to take on a Japanese mentality to serve the Empire. A large body of *Kominka* literature and fine arts was encouraged and produced to praise the war and the nobility of sacrifice, as well as the immense glories and privileges some Taiwanese had gained through participation in the "holy" warfare. Japan awarded collaborative writers with great recognition and honor. It became the indisputable high culture of the society against which people were measured for their merits. It was, in fact, the most influential institutional force in shaping the moral horizons of the Taiwanese people.

The *Kominka* movement, in effect, created a colonial society based on extreme "duty-morality" and hegemony. The premises of duty-morality were to obligate the Taiwanese to their masters without any reservation. The Taiwanese had neither self-esteem nor any meaningful political rights under this circumstance. But they were recognized and rewarded for adopting this duty-morality by their colonial master. The Japanese officials were supposed to be the examples to the Taiwanese, leading them to be dutifully faithful to the Emperor and the Empire. They were austere and despised the Taiwanese in general, but were committed and disciplined nationalists. It is fair to say that the Taiwanese were observing and practicing (not without any reluctance) some new codes of honor and virtue based on this colonial hierarchy from the late 1930s onward. The hierarchical and reciprocal relations between the colonial citizens and their rulers became "the way should be." This worldview was almost completely shattered on the eve of the "228 incident" in 1947.

It is estimated that from 1937 to 1944, the Japanese-speaking population increased from 38 to 71 percent.[30] To adopt Japanese style and morality, and to become a Japanese-speaking family gave some Taiwanese an illusion of being equal to the Japanese and perhaps superior to their Taiwanese neighbors. The resisting intellectuals in the past had asked the Taiwanese to "wake up" and resist colonial rule to regain their pride. The Japanese, however, soon succeeded in providing a "better" livelihood through modernization along with a new sense of social recognition and honor for ordinary Taiwanese through national education and institutionally encouraged identification with the Empire. All forms of resistance were almost completely silenced after 1935. The most respected Taiwanese leader Lin Xsiatang was also forced to take on some *Kominka* activities and an honorable position to collaborate with the Japanese, even though his reluctance and passivity were noticeable.

The assessment of *Kominka* and Japanese rule in Taiwan, in general, has been at the center of heated debate between the Chinese and the Taiwanese nationalists in recent time. The Taiwanese nationalists have been by and large sympathetic toward the situation the Taiwanese had experienced, and have refused to pass harsh judgments for past cooperation with the Japanese. Some even praise the Japanese, using the "good-old-days" to judge the corruption of the Chinese government that succeeded the Japanese in 1945. Some even suggested that the Chinese government and military were neither competent nor qualified to rule the Taiwanese because the latter had become more advanced and civilized.[31] On the other hand, the Chinese nationalists have strongly resented Japanese imperialism and Taiwanese condolence of their "lost past" under the Japanese. The fact that the Taiwanese could not see the evils the Japanese had done to the Taiwanese and the Chinese was simply abhorrent to them. They are often outraged if the Taiwanese should compare the "backwardness" of Chinese to the "modernity" of the Japanized Taiwan. They criticize Taiwanese nationalists for not knowing the suffering that the "Ancestor's Country" (*zhuguo*) had gone through because of Japanese imperialism. These differences in national identity, orientations about what is right or wrong regarding oneself, or good or bad regarding the past, always generate a lot of emotion on both sides in contemporary Taiwan.[32] Why are they so different? I will address this in the following section.

De-colonizing or Re-colonizing? The Tragedy of Unification

The Nanking government sent one of its most able generals Chen Yi to head the "take-over" government in Taiwan in October, 1945, three months after the Japanese surrendered. Taiwan was reunited with the Ancestor's Country again. And it is widely known that "the *er-er-pa shijian*" (or the "228 incident") that took place on February 28, 1947, marked the watershed of Taiwanese nationalism. The incident is actually better described as an "uprising" against the Chinese takeover government instead of being merely an "incident." The uprising was triggered by the arrest of a Taiwanese woman in a street of Taipei City for selling smuggled cigarettes, and the police shooting at a crowd who were attempting to protect the woman. The story spread out quickly and overnight the whole island was boiling with rage. Angry Taiwanese stormed police stations, government agencies and sought out and beat up mainlanders. Taiwanese elite and activists became united and organized after the uprising. They presented demands for stopping government corruption and abuses, and for autonomous government to Chen Yi and the central government in Nanking. The central government retaliated with military enforcement one month later. The subsequent crack down nearly caused the entire elimination of prewar Taiwanese leftists, and scores of death of Taiwanese activists and middle-class professionals who were suspected to support the uprising.[33] The survivors were silenced, put in prison, or forced into exile abroad.

Before the uprising, the Taiwanese welcomed the reunification with the Ancestor's Country though with a lot of uncertainty. The welcome was genuine for it meant the end of war and Japanese colonization.[34] The uncertainty about unification was partly caused by feelings of guilt for not resisting the Japanese and fighting for their own liberation, and partly by lack of knowledge about what would happen to them next. But after the suppression and the killings of Taiwanese following the uprising, people became very disillusioned and silenced out of fear. For many Taiwanese who escaped abroad becoming expatriates, Taiwanese nationalism entered a new era: now Taiwan must be an independent country. Taiwan should be for and of Taiwanese, and the Taiwanese for and of Taiwan. Dissident exiles first used Tokyo, and then later the United States as their home base to regroup and launch the present-day Taiwanese Independent Movement. Self-determination and national independence had been the political platform of the Taiwanese Communist Party in the late 1920s, but they had little chance to promote it among the public. Now, it became a popular idea among Taiwanese exiles around the world because of the tragedy surrounding the reunification and the rebellion. Though the exiles' movement was also divided into the Left and Right and other kinds of factional differences, overall the call for self-determination, for a categorical Taiwanese identity completely opposed to Chinese identity, was clearly conceived.[35]

This is not the place to discuss fully the reasons and background of the "228 incident." We can only focus on the identity issues central to this chapter. The differences between the Chinese and Taiwanese when they became united in 1945 were not outrightly racial or ethnic issues as argued by the contemporaries. Rather, the Taiwanese and Chinese were peoples with supposedly the same origin but forced to follow different historical trajectories, and subjected to different historical forces of self-transformation and identity formation.

We first have to realize that politically the Taiwanese were no longer traditional Han-Chinese in the 1940s. They had changed under colonization, through the national education and modernization promoted by the Japanese. The activists and their many followers had already acquired the necessary articulacy for self-definition and of a national community, named and imagined as Taiwan and Taiwanese. They had not succeeded during the Japanese period and were obligated to cooperate with the Japanese, but the notion of being Taiwanese who had the "right" to home rule was not erased, and the activists were ready to rise again after the War was over.

Though the Japanese Emperor surrendered and the Empire dissolved disgracefully, the underlying duty-morality of the citizens and government prevailed, and a sense of law and order persisted during the three-month transient period in Taiwan between the surrender of the Japanese and the official take-over by the Chinese. By and large, the Japanese were very depressed, scared, and waiting for deportation while the Taiwanese were celebrating though with a lot of uncertainty. But it was amazing that the Japanese did not suffer any retaliation from the Taiwanese. The Japanese government during the transition had allowed Taiwanese elites and civic associations to take more responsibilities of maintaining social order, and there was little sign of chaos and deterioration of social order despite the rapid down turn of the economy and the power vacuum.[36] This peaceful and cooperative relation after the War between the Taiwanese and the Japanese, or between the colonial subjects and their masters, has distinguished Taiwan from other places in the postcolonial period. Even today, President Lee of Taiwan believes that the Taiwanese might have been the friendliest people toward Japan in contemporary Asia.

Second, the Chinese themselves were no longer traditional Han-Chinese in 1945. Since the late nineteenth century, they had been involved in the transformation of the Chinese through nationalistic revolution. And the eight-year's Resistance-War against Japan had awakened the rather diversified and regional populations to the nationalistic and patriotic mission of *the* Chinese. The final victory in 1945 was sweet but bitter because millions of Chinese were killed, and its infant industry was almost completely destroyed during the War. Thus an extreme duty-morality of Chinese patriotism and nationalism was riding with the anti-Japanese sentiments. But this national transformation was taking place outside Taiwan. In sum, Taiwan and China had been on their respective paths of change in two different political fields with different moral horizons. Each had little contact with the other since the seceding of Taiwan to Japan in 1895.

The Chinese in general and the government did not know what to expect in Taiwan, and was certainly not ready to accept the possibility of the Taiwanese wish for home rule. Instead, a takeover government with a centralized administrative unit and exclusive and absolute power privileged by Chiang Kai-shek was set up in Taiwan, just like the former *Taiwan Sotoku* under Imperial Japan. In retrospect, the Taiwanese compared the new government to that of the Japanese, and found that the new government was just like the old colonial government, only more "backward," "poorer," and worse.[37]

Third, the Taiwanese bourgeoisie and middle class had their own aspirations for the reconstruction of Taiwan. This included not only the right to self-rule, but also normalization of the market and the economy from austere regulations during the War, and the replacement of Japanese with Taiwanese talents in many areas. They were inspired by the hope to ascend to a higher social status without being discriminated against by any outsiders. On the other hand, the Taiwanese leftists were also eager to educate and mobilize the underclass for both democracy and large-scale social reform. But they were having closer contacts with the Chinese communists instead of the Kuomintang. And very soon, they all realized that the Chinese takeover government did not recognize these aspirations at all.

In the beginning, the Chinese officials were thinking of building Taiwan as a "model province" for the rest of China. After all Taiwan had the infrastructure, from the education system to power plants that were built and left by the Japanese. In reality, however, this soon changed and Taiwan was used to support the reconstruction of the war-torn Ancestor's Country. Taiwan existed as a part of the larger national picture, not merely for itself. The Taiwanese were told that it was the time to "pay back" the Ancestor's Country. Shortly after, young Taiwanese men were drafted and Taiwan's economic resources were appropriated to support the civil war against communist insurgency on the Chinese mainland. This was equivalent of reimplementation of the extreme duty-morality, except that the new master was thought to be more abusive and corrupt. It was at this historical juncture that the Taiwanese and the Ancestor's Country diverged. The Taiwanese wanted to regain their "rights" to rebuild their own home and bring back their "normal" life, but the Chinese gave them more nationalistic "duties" and asked them to make more sacrifices. The respective moral horizons of the Chinese and Taiwanese were so different that they could not understand or respect each other despite their common origin.

Scholars have attributed the cause of the "228 incident" to factors including corruption, bureaucratic mismanagement, collapse of the economy (high unemployment, food scarcity, and high inflation), Chinese officials' nepotism, and so on. The incident followed by the persecution and killing of Taiwanese people are cited as the reasons that prompted Taiwanese nationalism. But critiques say that parallel incidents with similar backgrounds had also occurred in other former Japanese colonies in China, in cities such as Qingtao, Shanghai, and Harbin. Why should this particular tragedy become a watershed for national independence, when similar tragedies in China did not create other independent movements? The answer is not difficult to find if we look at the change and the transformation of the Taiwanese under the Japanese.

The de-colonization of the Taiwanese would imply replacing the legacy of Japanese high culture with proper Chinese codes. And, it would imply the replacement of Taiwanese moral horizons with more appropriate Chinese nationalistic ones. But this was carried out without the recognition and the incorporation of neither Taiwanese leftists nor the resistance elite. The Taiwanese were experiencing distrust and exclusion during this early period of reunification. As some of the Taiwanese tried to raise voices of disagreement or to criticize the abuses of Chinese officials, they were quickly criticized for being "poisoned by the mindset of being Japanese slave" (*riben nuhua sixiang*). The greatest irony was Taiwan, after the War, entering into another kind of humiliating situation through its reunification with the motherland. The Taiwanese were obligated to compromise their own integrity, to accept the inferior and new dutiful status assigned by the Ancestor's Country and its officials, who claimed to be the liberators. This kind of postcolonial situation sets Taiwan apart from the other major cities in mainland China.

Today many people are puzzled. They ask, how could the Ancestor's Country be foreign? And they argue: Chineseness was, and still is in the Taiwanese "blood." And "blood is always thicker than water." They do not understand the identity-transformations of both the Taiwanese and the Chinese since the twentieth century. Neither do they understand that the Taiwanese were deeply hurt by the way they were treated by the Chinese after the national unification. As Peng Mingmin, a prominent Taiwanese independence activist and the first presidential candidate representing the Democratic Progressive Party in 1996, recalled that after the "228 incident," his father (a physician practicing in southern Taiwan) felt so shameful for having Chinese "blood" (lineage) that he denounced his Chinese "blood." He even wanted his descendants to marry foreigners, so that they could never be Chinese again.[38]

The notion of a categorical Taiwanese in opposition to the Chinese was seeded in the minds of many members of this generation and their families. However, they could not speak their feelings out under the authoritarianism until the 1980s. This sentiment and morality became the basis for the Taiwanese Independence Movement in Japan and other places beyond the reach of the authoritarian rule of the Republic of China. Taiwanese feelings of being victimized by the mainlanders and the Ancestor's Country also became the underlying morality for Taiwanese mobilization inside Taiwan, later becoming a milder form of democratic movement.

National High Culture and the Authoritarian Rule

The Kuomintang government retreated to Taiwan in 1949, and one and a half million nationalistic mainlanders followed. The government was determined to maintain its claim to be the sole legitimate ruler of all China, which had been "stolen" by communist "bandits." It desperately needed patriotism and loyalty from both the Taiwanese and mainlanders for its mission to retake the Chinese mainland and for its own survival in Taiwan. The fundamental goal of nationalistic teaching in Taiwan was to educate all citizens, despite his or her origins, and instill loyalty to undertake this anticommunist and reunification mission. Mandarin and the reinvoking of traditional values and Confucianism became the dominant civic virtues. New citizen etiquette from proper manners to lifestyles was prescribed by Chiang Kai-shek personally and his *Chinese Cultural Renaissance Movement*.[39]

At the same time, not only was the legacy of the Japanese culture such as language and popular music disappearing very quickly in Taiwan, but so too were the Taiwanese dialects (of all sorts), traditional customs, folk religious practices, opera and music that were soon systematically defined as local, backward, superstitious, either harmful for national unification and/or national modernization. The Chinese official Mandarin traditions (of the ruling class) were upheld as the high culture of Taiwan. This was followed by "learning modernization values" from the United States and the rest of the "free world," associated with economic pragmatism in the years when the United States was Taiwan's strongest supporter, as a military ally and a "friendly" country that provided economic aid. The United States has replaced Japan as the country that the Taiwanese youth sought to receive postgraduate education since the late 1950s. The ruling class and nationalistic high culture were exemplified by upholding standard Mandarin as the means for education, official use, printing materials such as novels and newspapers, and also as the performance language of popular culture after television became popular in the 1960s. American English became the cosmopolitan language taught in the middle school (the seventh grade). As a consequence, a humiliating condition were created for many Taiwanese, of all dialect groups, for their language and accents, daily manners, tastes and styles, or more plainly put, for what they were, in the eyes of those from the high culture.

Charles Taylor argues that a strong unifying national high culture is likely to bring out a defensive reaction from minority groups, in the modern era when dignity and being recognized becomes a prevalent morality. And the elite will be especially aware of problems relating to respect and dignity in this imposed nationalistic assimilation.[40] This seems to be the situation in Taiwan since the 1970s. Ascending to its hegemony, the Mandarin–U.S. high culture experienced strong opposition from people of various origins in the late 1970s. People resisted being pushed backwards and began to reassert themselves as a reaction to this process. There have been many examples of this kind of resistance; for instance, (1) the *Home and Land Literature* (*Xiangtu Wenxue*) submitted by some leftists (who became pro-PRC nationalists later) and Taiwanese writers, in the early 1970s, (2) the aboriginal movement that has demanded de-stigmatization (known as the "Name Rectification Movement" in Taiwan) and the return of their traditional lands since 1984, and (3) the Hakka movement that has demanded recognition and the right to preserve their cultural identity and language, since 1989. These have been reactions against a nationalistic and nationalizing project that is seen as ultra-assimilationist in its objectives. Due to space constraints, I will focus on the identity issues surrounding *Taiyu* as an illustration of this kind of resistance against the nationalistic Chinese high culture.

The *Taiyu*-speaking population has been at least around 75 percent of the general population since the 1950s. But speaking *Taiyu* has been strongly discouraged and even punished by educators and public institutions because it is thought to be local, backward, and harmful for national unification. But for the same reason, speaking *Taiyu* in public, especially in front of mainlander officials, has become a defiant gesture since the 1980s. It has been an important vehicle in political rallies to arouse feelings and resentment against the high culture and the government in the past, and even today. I will explain how this came about in the following pages.

Though the Kuomintang government was at one time considered to be just an "outsider's regime" or another "colonial" regime by many

Taiwanese nationalists, it is indeed different from the Japanese. For instance, there were few barriers to those many Taiwanese attempting to obtain upward mobility through national education and economic growth. In a sense, other than setting up a national culture of domination, and suppressing local differences, the nationalistic ethos was genuinely "equal" in its attempt to make all people uniformly patriotic and loyal Chinese. Political persecution of dissidents were also "equal," in general, without any intentional bias against the Taiwanese; the same level of suppression was also used against dissident mainlanders in Taiwan, in particular against the leftists.

But the consequences of and reactions to this kind of "equal" domination varied among people of different origins. The majority of the Mandarin-speaking mainlanders did not find this pursuit of high culture to be a problem. They, in general, had a higher education level and had been part of that culture. They supported national unification and modernization with strong national duty-morality. Unless they were suspected to be associated with the communists or the leftists, or sought out by the right-wing regime for political differences, they were comfortable in sharing the nationalistic worldview. But it was humiliating for members from *Taiyu*-speaking and other groups who could not master the language or the nationalistic etiquette as easily as the mainlanders. In general, the *Taiyu*-speaking population had a lower level of education, and were constantly frustrated and felt humiliated just for being themselves. An internal division between the two peoples was thus created, despite attempts at nationalistic assimilation.

On the other hand, since the Kuomintang regime allowed some degree of political participation for the Taiwanese, in the late 1970s, the postwar and post-228 generation of Taiwanese activists began to reemerge from the previous silent period. Eventually, the Taiwanese have been able to develop a "right-morality" as opposed to just "duty-morality" through large-scale political mobilization. Different from the "duty-morality", the "right-morality" argues for self-rule and political and civic rights, including the right to form an opposition party, to have freedom of expression, and the right to demand an end to martial law rule, to lift the policy of checking whether citizens thought—and acted in the proper way, to revise the militaristic tendency found in national education, and to allow the use of *Taiyu* in public. It would be fair to argue that the democratic movement in Taiwan since the 1980s has meant self-integrity and self-respect, much more so for the Taiwanese than for the mainlanders. Besides, the movement also put the mainlanders on the defensive for their support for the national (Mandarin-centered) high culture.

Under authoritarianism, younger and newly educated Taiwanese activists were able to articulate their discontent and rights as citizens against the government, during what is generally known today as "election holidays." The "holidays" are periods during which the regime did not exercise the stern suppression it normally would have during ordinary (or the non-election) times.[41]

Although neither *Taiyu* nor Hakka ever developed widely accepted writing-forms, the hurdle for a writing language to reach the general public was overcome completely in the late 1970s, when the Chinese literacy rate went over 90 percent. The opposition activists could publish dissident magazines in Chinese, and play a mouse-and-cat game (sometimes at a very high personal price) with the "thought police" of the Garrison Command by printing out their ideas and critiques.[42] These publications paved the way to mobilize the Taiwanese public against the pressure of assimilation from the high culture circles and the very biased mass-media influenced by the Kuomintang. The pervasive feeling of unfairness or frustration among the postwar generation of Taiwanese was generally articulated by the opposition under these adversarial conditions: (1) the authoritarian rule which suppressed people's liberty and basic human rights, (2) the imposition of official high culture, which was exemplified by the exclusive use of Mandarin in public, (3) the creation of a culturally advantaged mainlander group vis-à-vis the less advantaged Taiwanese group, (4) the opportunity to use "election holidays" to use *Taiyu* in political rallies, and (5) the publishing and spreading of dissident ideas to the general public. The cultural gap between the opposition elite and the populace was greatly narrowed after the 1970s. A larger consensus of the meaning of being Taiwanese, the more established "natives," was constructed in this period through the expansion of political participation by the general public and the attribution of suppression to "outsiders" regimes dominated by the mainlander group. Their efforts have helped to crystallize what we know today as the "ethnic politics" in Taiwan, the politics of the mainlanders versus the Taiwanese.

The Pragmatic Aspect of Taiwanese Identity

There have been many studies of the current development in "ethnic politics," in Taiwan. I will not repeat their findings here. It has been argued that the Taiwanese are much more likely to support the Democratic Progressive Party and Taiwan independence, whereas the mainlanders are much more likely to support unification, and the nationalistic

Chinese New Party and the Kuomintang. This chapter is different from these studies because I push for the understanding of what it means to be Taiwanese in history, and the changes in its meaning. I have also focused on the moral aspects of national identity, the role of high culture, and the importance of recognition and respect. The general theme I have argued so far is that the recent meaning of Taiwanese as a national group did not exist until the 1920s, and its current meaning as having the "right" to be distinctively Taiwanese did not become popular until the democratic movement between the late 1970s and early 1980s.

Nevertheless, Taiwanese nationalism seems to have taken another significant turn during the 1990s. Since then many Taiwanese have chosen to side with the Kuomintang, and support the institutional arrangement of the ROC. The name and the symbols of the previous "outsiders' regime" continue to exist, and many people still vote for the Kuomintang despite the rising tide of Taiwanese nationalistic thinking. Supporters of Taiwanese nationalism have explained this in terms of the "indoctrination" of nationalistic education and ideology. They argue that the Taiwanese people were not awakened to the fate of an "authentic" Taiwanese (the expression of "*jiang-gang Tai-wan lang*" in *Taiyu*). But the success of indoctrination cannot explain why even after political liberalization the support for unequivocal Taiwanese independence has not exceeded a low 15 percent margin. And it cannot explain why the DPP, the champion of Taiwanese independence since it incorporated "people's self-determination" in its party platform in 1991, has so far never won over 30 percent of the votes in various kinds of national elections. Its support was even lower than 20 percent when longtime Taiwanese dissident Peng Mingmin represented the party as presidential candidate in 1996. And, neither can it explain the softening of the DPP on issues of Taiwanese independence since 1994, its turning to the "middle-of-the-road"; an act that provoked the spinning-off of the more militant Taiwanese Nation Building Party (*jianguodang*).[43]

This later development of Taiwanese nationalism has moved beyond the call for "dignity" and "recognition" from the mainlander's regime. The most significant phenomenon has been the "indigenization" of the Kuomintang, or the change in the ruling elite from mainlanders to Taiwanese. The pattern of power sharing within the Kuomintang under its current leader Li Denghui changed significantly after 1989. Lee successfully promoted himself and his supporters of Taiwanese origin from the marginal to the central inside the Kuomintang. Sometimes he made use of the support of Taiwanese nationalistic ideas and of active assistance of DPP members when he attacked his mainlander enemies in the KMT. He also crafted the basic tone of nationalistic rhetoric by asserting concepts that have abstract and vague meanings such as "Community of Fate," "Taiwan first," "new Taiwanese," and of course, the "two-State theory." He has also discredited the Republic of China as an "outsider's regime"; the new era of the Kuomintang started from his ascension as its indisputable leader in 1992.[44] This "indigenization" of Kuomintang is not only a source of pride and dignity for the Taiwanese who used to work for the Kuomintang, but also a legitimization process for the status quo interests and the ruling party. Consequently, by promoting the Taiwanese identity and toeing the line of the mainlander elite inside the party, the Kuomintang was able to sustain itself in the post-authoritarian era.

We must know that politics is also a form of action for the extraction and redistribution of economic resources and material well being. And there is always a potential for loss in pursuing hasty and complete Taiwanese independence, regardless of its alleged nobility associated with self-determination. When under authoritarian pressure, the potential loss could even mean death sentence or long prison-time for dissidents. Therefore, except for Taiwanese expatriates, most Taiwanese activist movement, in Taiwan did not, in effect, advocate Taiwan independence, not to mention ordinary Taiwanese before the political liberalization between 1986 and 1987. After liberalization and the final assurance of freedom of expression in 1991, the potential personal loss was about economic and other utilitarian gains. The Kuomintang was still the ruling party at the time when this chapter was first written. It had the advantage of the incumbent in administration, and in the daily management of extracting and redistributing resources. It was and still is one of the largest conglomerates in Taiwan, wielding enormous economic clout. Though the majority of people felt skeptical toward Kuomintang's overwhelming economic resources, the party could and still can play with people's needs for economic gains and manageable social and political changes with relatively moderate rhetoric, thereby maintaining the status quo by rejecting an outright Taiwanese independence.

Another factor that prevents the Taiwanese from pursuing an all-out nationalistic political position has been the PRC since 1995. The PRC had repeatedly criticized the tendency of Li and Taiwan for moving dangerously toward Taiwan independence with hostile words and intense military threats, such as testing ballistic missiles near Taiwan's waters, or holding large-scale amphibious exercises along its eastern coast. It can be said to be the first time that the Taiwanese nationalism faced the immediate danger of having to confront a Chinese invasion since the "228 incident" in 1947. The nature of the risk of Taiwanese independence is thus shifting from provoking internal unrest between the Taiwanese and mainlanders to provoking a showdown between Taiwan as a whole and the PRC.

Many recent opinion polls have indicated that the bullying attitude of the PRC has actually increased the consciousness of being Taiwanese, the morality of self-rule, and self-constraint. The price of abandoning this symbolic Chinese nation-ness completely in current political institutions would include not only stirring up resistance from mainlander Chinese in Taiwan, but also, and more seriously, could provoke invasion from the PRC.

At the same time, the increasing economic exchange and personal visits between the two places since 1989, has further complicated nationalistic politics. To the displeasure of many Taiwanese politicians, Taiwan's economic growth has increased its dependence on investment in and trade with China recently. The negative impacts on Taiwan's economy because of the instability of cross-strait relations and domestic politics have been observed and widely debated. Taiwanese people are not unaware of the danger of pursuing Taiwanese nationalism too far. Even different presidential candidates of the DPP have shown their understanding of the consequences. The general opinion stated by the DPP and its presidential candidate Chen Shuibian, in the election year of 2000, was that Taiwan has been an independent country (or state) already, and thus there is no need to declare independence, and, they said, the name of this country is the Republic of China.[45]

From another point of view, the dignity of the Taiwanese also comes from the improvement in their economic conditions and material well being in the last fifty years or so. Contemporary narratives of being a Taiwanese are also about Taiwan's economic success. In spite of many internal differences, to preserve these economic gains and the democratic rights in Taiwan, make everyone necessarily have stakes in the status quo. There is no doubt that people in Taiwan may have very strong and different opinions about their own identities. Some are ethnic-cultural while some are national-political. However, the democratic political institution of the ROC, regardless of its infancy and awkwardness, seems to hold the differences together successfully so far. It has given people room to work out their strong feelings and dignity issues originating from the past, within its framework.

It is arguably true that since the late 1990s, the most agreed-on collective belief in Taiwan is neither Taiwanese independence nor Chinese nationalism. It is, rather, the belief in the importance and the "good" of continuous economic prosperity and development. Prosperity is not completely independent from the problem of pursuing a categorical and a respectful Taiwanese identity, since economic success and modernization have been significant in the story of being Taiwanese. Thus, to ask people to give up economic prosperity or to endanger them as a price for the independence of Taiwan is not a popular idea among the general public. The people in Taiwan are indeed very vocal about their differences, and with very conflicting identity attachments, but they have been pragmatic and rational, in general, in pursuing their nationalistic claims. The legacy of the political institutions of the ROC, thus, can remain with incremental reforms for consolidation of democracy and self-rule.

Some Concluding Remarks

I have tried, in this chapter, to present the zigzag course of the development of Taiwanese national identity. The story being told about a distinctive Taiwanese nation goes back to at least the 1920s. But it did not evolve into a full-fledged nationalistic movement overnight. The painful and lingering memories of the "228 incident" in 1947, and the overwhelming pressure for assimilation and the opposition movement against authoritarianism since the 1970s, have helped to crystallize the meaning of "authentic" Taiwanese in opposition to the Nationalist Chinese. When the reunification of Taiwan and the Ancestor's Country took place in 1945, the Taiwanese and the Nationalistic Chinese were actually standing on different historical trajectories that were forcefully created by imperialism from the outside. They became different "peoples" consequently. The identity stories continued to be different for them after the 1950s. The imposition of the Chinese high culture has made many Taiwanese feel "humiliated" and led to the creation of an "ethnic" division between the mainlander Chinese and the aiwanese. But what really matters are their respective identity stories and their different moral horizons, which have been in opposition to each other in the past. It is acceptable to think of them as two different ethno-national groups, as today we find them defining themselves and also believing in themselves. But it would be wrong to think that their differences were in fact caused by *a priori* ethno-national differences.

Respect, dignity, and recognition are important as we seek to understand the moral aspects of identity formation in history. But "real" politics is also about gains and losses. The indigenization of the Kuomintang since the 1990s has helped the existing political institution of the ROC continue to serve as the only viable framework that can hold all the differences together. This is possible only when many people also believe in the importance of political stability and national security for economic prosperity, civil liberty, and freedom, not only in terms of the absolute value of national dignity, which can generate strong nationalistic fervor.

Nowadays, Peking officials tend to make a mistake insisting that Taiwan needs to be reunified with China, the Ancestor's Country. It is ironic that Taiwan has already been unified with China since 1945. And the unification has made many Taiwanese pay a great price. The legacy of the political institution of the ROC and the political discords between the Mandarin-speaking mainlander and the *Taiyu*-speaking population can testify to the existence of this "unification" reality. Peking officials sometimes make another mistake by saying "Taiwan and China have been separated from each other for about fifty years." From the Taiwanese perspective, Taiwan and the rest of China have been, in fact, separated for more that one hundred years. Indeed, a lot of significant events and historical happenings have impacted profoundly on the common folks in both mainland and Taiwan without the participation of the others. The identity differences need to be recognized and respected, not only for those between the mainlanders and the Taiwanese in Taiwan, but also for those between the peoples in Taiwan and in mainland China. Not knowing properly the historical and fundamental sources of these differences can only lead to extreme patriotism and jingoism. In the current situation, any pressuring tactics toward hasty independence or unification is bound to be disastrous for all involved.

Notes

A previous version of this chapter was published in China Perspectives no. 28, March–April 2000.

1. For instance, Ernest Gellner's *Nations and Nationalism*. Ithaca: Cornell University Press, 1983, and Eric Hobsbawm's *Nations and Nationalism since 1780: Programs, Myth, Reality*. Cambridge: Cambridge University Press, 1990, are the two prominent books holding this position
2. As Eric Hobsbawm has said about the ideological nature of nationalism ". . . that no serious historian of nations and nationalism can be a committed political nationalist . . . nationalism requires too much belief in what is patently not so." Hobsbawm, ibid., p. 12.
3. Charles Taylor, *Sources of the Self, the Making of the Modern Identity*. Cambridge: Harvard University Press, 1989, p. 35.
4. Ibid, p. 27.
5. Ibid, p. 51.

6. Anthropologists believe that Taiwan aborigines are members of the Austronesian family.
7. Li Guoqi, "Qingji taiwan de zhengzhi xiandaihua-kaishan fufan yu jiansheng (1875–1894)" (The Political Modernization of Taiwan in Qing Dynasty—Open Mountain Territories, Cultivating Savages and Establishing the Province) in *Zhonghua wenhua fuxing yuekan*, vol. 8, no. 12, 1975, pp. 4–16.
8. Chen Qinan "Tuzhuhua yu neidihua: luen qingdai taiwan hanren shehui de fazhan." (Become Native and Become Inland: On the Development of the Society of the Taiwan Han Folks). In *Zhongguo haiyang fazhan luenwenji*. Taipei: Institute of the People's Three Principles, Academia Sinica, 1984, pp. 335–366.
9. As one of the most famous Chinese nationalistic intellectuals during the early twentieth century, Liang Qichao, once put it, it was only after the loss of this first Sino-Japan War and the signing of the *Shimonoseki Treaty* that Chinese intellectuals became finally awakened by their national crisis.
10. About 14,000 Taiwanese were killed in the first year of Japanese advancement in Taiwan. And about 12,000 were killed from 1898 to 1902. See Huang Zhaotang, *Taiwan na xiang na li si wen (Taiwan Nationalism)*. Taipei: Qianwei, 1998, p. 9.
11. See Wu Zhulin's widely acclaimed novel, *Yasiya de guer (The Orphan of Asia)*.
12. Luo Fuxing was a Hakka who became involved with the Kuoimintang and Hakka nationalistic Qiu Fengjia. His plan was to organize a rebel to drive out the Japanese. This incident sent more than one thousand Taiwanese to execution by Japanese police.
13. See Wang Yude (or Ong Jok-tik) *Tai-wan-kumen de lishi (The Suffocating History of Taiwan)*. Taipei: Qianwei, 1999, p. 129.
14. Huang, *Taiwan na xiang na li si wen*, p. 13.
15. Huang, ibid., p. 14.
16. Wang, *Tai-wan- kumen de lishi*, p. 132.
17. He has been the Secretary of the Propaganda of the DPP in the early 1990s, and now is a college professor of Taiwanese history.
18. Chen Fangming, "zhimindi geming yu taiwan minzu luen- taiwan gonchandang de 1928 nian gangling yu 1931 nian gangling" (Revolution in Colony and Taiwan Nationalism, the 1928's and 1931's Outlines of Taiwan Communist Party). In Shi Zhengfeng, ed., *Taiwan minzu zhuyi*. Taipei: Qianwei, 1994, pp. 287–320.
19. The left-wing activists feared that a representative Taiwan Council, supposed to be elected indirectly from regional elite, would be dominated and manipulated by the elite's interests, whereas the poor masses would still left exploited by the landlords.
20. Huang, *Taiwan na xiang na li si wen*, pp. 49–50.
21. Taiwan minzhong dang (the People's Party) was the first political party formed by the Taiwanese and Dr. Jiang weishui in 1927. It was closed down by the Japanese in 1932. Jiang weishui was inspired by the southern revolution and the northern expedition led by the Kuomintang in China. *Minzhong dang*'s party flag actually resembles that of the Kuomintang's.

22. See Douglas Fix, *Taiwanese Nationalism and its Late Colonial Context.* Ann Arbor: UMI Dissertation Service, 1993, pp. 76–77.
23. Huang *Taiwan na xiang na li si wen,* (note 10) p. 139
24. For instance, see Wang, *Tai-wan- kumen de lishi*, pp. 127–138.
25. Huang, *Taiwan na xiang na li si wen*, pp. 48–50.
26. Huang, ibid., pp.10–11, 17.
27. But for these last measures, especially giving up ancestor's tablet for the Emperor, many had drawn the line by refusing to do so or by doing so only under force.
28. Huang, *Taiwan na xiang na li si wen*, p. 17.
29. The total number of Taiwanese in the Japanese military was about two hundred thousands or more. During the last year of the War, more than forty thousand Taiwanese were also drafted (involuntarily) to join the military.
30. Wang, *Tai-wan- kumen de lishi*, p. 148.
31. For instance, both Huang Zhaotang and Wang Yude have refused to pass easy judgment on Taiwanese collaborators in their books.
32. And this difference in people's moral horizons, in effect, has contributed to the repercussion, inside the Chinese Nationalists and Kuomintang in 1994, after the interview of President Lee by Shiba Ryotaro, mentioned earlier. Lee was criticized by the Nationalistic Chinese for sympathizing with the *Kominka* movement and Japan's rule in Taiwan, and hence supporting the opposition; on the other hand, Lee was also highly praised by many Taiwanese for voicing their feelings.
33. The estimates of the killings of Taiwanese range from 8,000 to 22,000, depending on the sources. The entire population of Taiwan at that time was about six million.
34. Some argued that the postwar Independence movement idea began from the proposal made by a small number of mid-ranking Japanese officers and collaborating Taiwanese elite during the first few weeks of Japanese surrender (see, for example, Hsu Jielin, *Taiwan zhanhou shiji*, or *The Historical Notes on Post-War Taiwan*, vol. 1, ch. 7, and the net edition at http://aff.law.ntu.edu.tw/after_war/content.html,12/10/99). But it was just a bubble conspiracy with support from neither the Japanese high official nor the general public. Its importance should not be over-emphasized.
35. Huang, *Taiwan na xiang na li si wen*, pp. 17–18, 74–76.
36. See Douglas Fix, *Taiwanese Nationalism*, chapter 5.
37. Wang, *Tai-wan- kumen de lishi*, (note 13) p. 157. Dr. Wang lost his elder brother during the 228 incident when he was still a student. He was forced into exile to Japan in 1949. He commented angrily in his book: "only now (referring to the abusiveness of Chinese officials in Taiwan), Taiwanese began to miss the Japanese period. Taiwanese despised Japanese and had called them 'dogs.' 'Dogs' bark, but 'dogs' will watch the door for you. Chinese are 'pigs.' 'Pigs' have no worth except for glutting themselves."
38. Peng Mingmin "*Peng mingmin huiyilu—the Taste of Freedom.*" Taipei: Qianwei Publishing Co., 1988, p. 80.

39. Allen Chun, "From Nationalism to Nationalizing: Cultural Imagination and State Formation in Postwar Taiwan," in Jonathan Unger, *Chinese Nationalism*. New York: M. E.Sharpe, 1996, pp. 126–147.
40. Charles Taylor, "Nationalism and Modernity." In Robert McKim and Jeff McMahan, *The Morality of Nationalism*. Oxford: Oxford University Press, 1997, pp. 31–55.
41. The government thought it had many other ways to win elections despite a small number of "agitators." It might have been under pressure of American human rights' foreign policy since Jimmy Carter. The "holiday" was therefore allowed to woo recognition of the legitimacy of the elections.
42. An extreme case was the suicidal protest by Zheng Nanrong who set himself on fire in 1989 when the Garrison Command attempted to arrest him for printing the Draft of the New Constitution of the Republic of Taiwan in his magazine, *Ziyoushidai* (Time for Freedom).
43. For DPP's crisis during self-transforming, from a champion of Taiwan independence to a more moderate political party, see Guo Zhengliang, *Minjindang zhuanxing zhi tong* (The Pains of DPP's Transformation). Taipei: Tianxia, 1998.
44. This can also be found in the same interview called "the Sadness of Being Taiwanese" reported by Shiba Ryotaro in 1994.
45. On this matter, the DPP and President Lee concur with each other after the submission of the "Two-States Position" by Lee in 1999.

Chapter 3
Shifting National Identities in Public Spheres: A Cultural Account of Political Transformation in Taiwan

Duujian Tsai

> In the past, we were not allowed to speak Taiwanese in public . . . So far, those ruling regimes in Taiwan have all been alien ones. As of late, I am no longer afraid to say so. Kuomintang [KMT, the current ruling party in Taiwan] is also one of these alien regimes. It is a political party to govern Taiwanese. In consequence, it is necessary to transform this party into a Taiwanese one. For years, Taiwanese over seventy, like me, were unable to live without nightmares. I must never let our children or grandchildren suffer the similar experiences . . . Modifying the constitution, accomplishing democratic transformation, implementing democratic rules, and, finally, letting people elect their own president will fulfill the expectations of Taiwanese.
>
> Li Deng-hui "The Sorrow of Being a Taiwanese" (1994)

In May 1994, the *Independence Evening Post* in Taiwan translated an interview between Taiwanese President Li Deng-hui and Japanese journalist Sima Liutaole. This interview, entitled "The Sorrow of Being a Taiwanese", signaled a new age of consolidation between the alien regime and Taiwanese nationalist movements. However, this new era of consolidation also includes a new style of confrontation between people and the Government. While President Li's voice pacified the strongest Taiwanese nationalist elites, his administration took new steps to censor Taiwan's popular public spheres, through attacking underground radio broadcasting stations. These radio broadcasting stations had played significant roles in pushing political reform in the past. In Taipei, such official attacks initiated the most extensive and bloody protest in ten years.[1] At the time of political reconciliation between the ruling party and the opposition elites, these mass riots presented a puzzle regarding Li's democratic reform and his promise to meet the expectations of all Taiwanese people (Li 1994). To solve this puzzle, this research asks: Who are the "Taiwanese"? How has Taiwanese national identity emerged in various public spheres? What are the links between Taiwanese national identity and Taiwan's democratic reform?

To address these questions this research will focus on three moments, the Siang-tu wen-siue Debates, the Taiwanese Consciousness Debates, and the political discourse since the Sie Li-fa Debates, as well as their historical contexts to ascertain the emergence of Taiwanese national identities in the 1980s. These historical events have been selected because they have: (1) received wide attention in public spheres, specifically in the major mass-media, at the time of occurrence, (2) presented the competing forms and contents of national identities, and (3) been supported by networks of intellectuals possessing significant power to support their viewpoints in public spheres at those moments.

In discussing these three debates, I will also address the following hermeneutical issues: How can cultural studies contribute to a sociological understanding of a non-Western society? How can both cultural studies and postcolonial studies provide a new explanation of a non-Western country as well as contribute to creating an independent subjectivity/historicity of that country and shape a vision for its own future?

To provide an appropriate sociological account of non-Western subjectivity/historicity, this research project has borrowed from cultural studies in three ways. First, being self-reflexive to its political implications, this project endeavors to provide sympathetic yet critical reflections on the recent past of the Taiwanese political transformation. Second, it adopts an interdisciplinary approach to ascertain historical dynamics through the lens of cultural formation, that is, the power shifts of emerging, dominant, and residual hegemonies. Third, this research represents an attempt to understand the current state of Taiwan's politics with possible visions for its further transformation. In so doing, this research project provides a more or less self-sufficient account of Taiwan's most recent history and its future. The goal of this paper is to free understandings of Taiwan's historical process from those Western-oriented theories of modernization. In the past, Taiwan's social and political changes have been examined by using modernization theory or dependence theory. As a result, economic development, especially following the path of Western countries, becomes the most important foundation for political and social development, and in particular, democratization. By providing an account of the ways in which Taiwanese created their own different identities, this chapter is not going to define any identity beforehand; instead, it will search the meanings of various identities in historical texts and contexts.

More precisely, this research has adopted a modified discourse analysis inquiring into Taiwan's contemporary political transformation during the 1980s. During this transformation, concern with the Taiwanese national identities did not appear in the major mass-media until the very end of this period. This research, following Foucault's idea of the history of the present, is to ascertain the beginning of newly emerging national identities in public spheres.

This research will examine the emergence of Taiwanese national identities according to the formation of a new discourse in Taiwan: ideas of national identities and articulations of state building. The central theme of this chapter is Taiwanese national identities. In the following section, I will first use Wachman's recent work on Taiwanese national identity to illustrate current problems in Taiwan studies. Thereafter I will explain my use of Chatterjee's two analytic angles, the thematic and the problematic, in exploring the history of Taiwanese national identities. I will also discuss how current literature on national identities and state theory may create a new set of explanatory themes for this research.

Present Problems in Taiwan Studies

In 1994, Alan M. Wachman, a political scientist, published a book discussing Taiwan's democratization and national identities. He argues that the most recent democratic political reform (from 1987 to 1992) resulted from a contested interplay between a Chinese national identity and Taiwanese national identity. He views both the dominant party, the Kuo Min Tang, and the opposition party, the Democratic Progressive Party, as representing Chinese and Taiwanese national identities, respectively. He claims that popular desires for stability and economic growth as well as the strategic calculations of the political elites of both parties have produced incremental and restrained reform, leading to an evolutionary path for democratization. Wachman's arguments on the cultural perspective of national identities and the political perspective of democratization contribute an important new understanding to Taiwan studies, which in the past was mainly limited by politicoeconomic perspectives. However, Wachman's research did not explore how Taiwanese national identities emerged in public spheres and gained sufficient strength to contest the so-called dominant Chinese national identities. This chapter, therefore, will address this concern, but will first provide a critical reflection on the limitations of Wachman's research.

Wachman's research has both theoretical and empirical limitations. Theoretically, Wachman situated his arguments in politicoeconomic-oriented rational choice theory. He shifts his explanation about Taiwan's social and political change to a teleological path of politicoeconomic development. More precisely, Huntington's modernization metanarrative becomes Wachman's theoretical plot to reappropriate Taiwan's most recent history. As a result, cultural schemas become indifferent to political and social change. That is, all actors, despite their national identities, will calculate their actions merely based upon Taiwan's political stability and economic development. Empirically, Wachman considers the KMT and DPP to represent the Chinese and Taiwanese national identities, respectively, without paying sufficient attention to meaning shifts in these identities. Consequently, he fails to explain crucial events, such as the national identity conflicts within the KMT, the establishment of the New Party by some of the KMT old guard, and the cooperation between President Li and the DPP to help a Taiwanese Premier, Lian Jhan in power. To go beyond these limitations, we have to identify inappropriate assumptions in Wachman's research.

Three main limitations can be identified: the rigid definition of national identities, a simplified interpretation of democracy, and non-reflexive Western centralism. In defining national identities, Wachman merely gives a functional account of national identities, and views national identities as an integrating mechanism for a political party. He also sees political parties with different national identities as having opposing political goals. Such a polemic vision of national identities is not appropriate to illustrate the complexities of political struggles. As pointed out by Stoler and Cooper, "Political possibilities do not just lie in grand oppositions, but in the interstices of power structure, in the intersection of particular agenda, in the subtle shifting of ideological ground" (1996: 13). To portray such opportunities, Wachman should have provided a more flexible vision of national identities.

As for his simplified interpretation of democracy, Wachman views democratization merely as an inclusive institutional reform process in state building. He fails to note that liberal democracy is, in fact, an inseparable part of the exclusive nature of nationalism in the modern history of either Western or third-world countries. Such evidence has been well articulated in colonial/postcolonial studies. As Mehta identified, liberalism is grounded in a specific set of cultural norms referring to a constellation of social practices riddled with a hierarchical and exclusive density (1990). Thus, as Pender stated, exclusive cultural practices were, in fact, justified by the economic interests of the empire in the age of liberalism. For Stoler, "Inclusionary law left ample room for an implementation based on exclusionary principles and practices" (1992). As a result, democracy should not be merely assumed through the form taken by state institutions; its exclusive nature should also be explored through social and cultural practices.

As for his non-reflexive Western centralism, Wachman applies Hungtington's theory to explain Taiwan's historical experiences. However, he fails to acknowledge the possible negative effects of Western democratization on Taiwan, and ignores the role of Western power in the process of Taiwan's democratization. More important, Wachman criticizes the KMT unwittingly for only granting citizenship to residents in Taiwan rather than to all people in China, and sees the party as being inconsistent with its One China policy (1994: 259). Such arguments exclude certain cultural and social data as critical from other vantage points.

To avoid Wachman's limitations and provide a more sufficient account of the subjectivity/historicity of a non-Western case, my research will provide theoretical angles to analyze local texts. It will give special attention to perspectives on globalization and localization. More precisely, this research will pay particular attention to the universal dimensions and local particularity of various national identities in Taiwan. As pointed out by Hobsbawm and Ranger, Anderson, and Chatterjee, nationalism originates in Western modernity, but is modified by local tradition and historical memory (Anderson 1991; Chatterjee 1986; Hobsbawn and Ranger 1983). These two dimensions are theorized by Chatterjee as thematic and problematic (1986). Thematically, every country must form a more or less homogeneous national culture based upon post-enlightenment ideology, that is, subjectivity, progressive, and logicism. Problematically, a sense of "imaged community" has to be constructed by reinventing the historical tradition and collective memory particularly in social and cultural contexts. The thematic and the problematic are the two theoretical angles for analyzing national identities during this time period.

While we ascertain various national identities in Taiwan, it is also important to understand their relationship to state building. Benedict Anderson argues that the term "nation-state" represents a convenient shorthand expression, but that we tend to thereby "overlook the fact that a tiny hyphen links two very different entities with distinct histories, constituents, and 'interests'" (1990: 94). In exploring the most recent political change in Taiwan, we must recognize the influences of titanic conflicts between national identities, as extrastate nativist solidarity movements confront an alien state from China (Gold 1986). While viewing the emergence of Taiwanese national identities as a new cultural formation, we have discovered the state and civil society articulating competing forms of social integration. Consequently, we must relate differences in discourses on state building to shifting meanings in national identities.

This chapter will examine the discourses underlying state building in Taiwan and discover their interactions with national identities. The state will not be viewed as a functional component of the whole society, a capitalist instrument to consolidate dominant economic relations, a rational actor concerned with its own profit, nor a homogeneous structure for exercising power. Rather, my research views the state apparatus as a strategically selective terrain, which has asymmetrical effects on the organization of power. Furthermore, within these strategically selective limits, the actual outcome of state power depends on the changing balance of forces engaged in political action both within and beyond the state (Joseb 1990: 303). The state, therefore, is viewed as a discursive field with its power relations and boundaries shifting over time. As a result, the idea of state building will change accordingly and will be ascertained through an analysis of visions about this process.

Finally, rather than following Wachman's appropriation of Taiwan's history by a Western metanarrative, this research will attempt to contribute to a new understanding of non-Western subjectivity. Moreover, while inquiring into the conflicts between people possessing different national identities, my research learns from colonial/postcolonial studies that "colonialism did not exist in the singular, but in a plurality of forms and forces—its particular character being shaped as much by political, social, and ideological contexts among the colonizers as by the encounter with the colonized" (Cromaroff 1985: 680–681). Consequently, this research will see national identities as existing cultural schemas that are subject to political appropriation by both rulers and ruled. Shifts in power relations and meanings of these national identities will be explored through the lens of emerging, dominant, and residual hegemonies (Williams 1977). To this end, a modified discourse analysis will be applied to exploring the emergence of Taiwanese national identities in the following three dimensions of the formation of a new governing discourse in Taiwan: ideas of national identities, articulations for state building, and visions transforming the existing ethnic relations to new social formations.

Methodology

The modified discourse analysis of this research includes a post-structuralist text analysis with an interpretative account of meaning shifts in defining national identities. It will be set within a Foucauldian discourse analysis focused upon an analytic account of power/knowledge. Discourse is defined as a group of ideas, their supporting institutional and personal networks, and their performative practices. These ideas are detailed in individual texts or articulations. A post-structuralist reading of these texts/articulations will problematize the universality of a concept (signifier) and explore its meanings (signified), referring to a set of situational references in a particular context. These texts/articulations related to a particular concept may be seen as a statement. Following Foucault's approach, this research will "grasp the statement in the exact specificity of its occurrence; determine its conditions of existence, fix at least its limits, establish its correlation with other statements that may be connected with it, and show what other forms of statement it excludes" (1976: 28). In so doing, national identity, a derivative Western concept, will be interpreted differently in a non-Western local context. Therefore, not only can a non-Western subjectivity/historicity be theorized, but the meaning shift of Western categories, such as nation, can also be explored.[2] Along this line, the modified discourse analysis will give a more sufficient account of the meaning shift of a concept in a context of asymmetric power relations.

To this end, texts emerging three incidents of debate on national identity, the Siang-tu wen-siue Debates, the Taiwanese Consciousness Debates, and the Sie Li-fa Debates, will be the main focus of this paper. However, this research will not merely focus on the texts of the aforementioned debates, but will also pay attention to their situational contexts. At the same time, this research will not assume an intrinsic identity for each historical agent of the various national identities. Rather, it adopts Homi K. Bhabha's view of nation as narration, wherein there is a split between the continuous, accumulative temporality of the pedagogical, and the repetitious, recursive strategy of the performative (1990: 297). Bhabha argues that the subject of a national discourse—the social actor—is in a state of discursive ambivalence that emerges in the contestation of narrative authority between the pedagogical and the performative (299). As a result, emerging, dominant, or residual hegemonies are pedagogical cultural schemas available for social actors, and their meanings can be shaped and reshaped in each performative practice. A hybrid form of people's national identities may also be presented by appropriating these cultural schemas in making "distinctions" between the "we" community and the "other." Hence, the meaning of national identities could not be asserted before public debates and would be better explored retrospectively thereafter. This research will therefore first briefly narrate an event and then identify what it tells us about related national identities. This approach will contrast the Taiwanese national identities with other competing national identities.[3] Comparisons between these events will also be made to present both continuities and the changeable nature of national identities.

By incorporating a Foucauldian discourse analysis with post-structuralist text analysis, this research will not only be able to focus on the three aforementioned debates, but also to portray these debates as part of the discursive formation of Taiwanese national identities during this time period. However, it also differs from Foucault and Said's approach, which merely focuses on particular Western governing discourse, by viewing discourse on the Taiwanese national identities in relational terms with Williams' ideas about hegemony (Foucault 1977; Said 1988; Williams 1977). Such approach allows analysis of power relations between the Taiwanese national identities and other competing national identities. Moreover, while identifying shifting dominant hegemonies, this paper will analyze the representation of such hegemonies in intellectual networks, using sources from Taiwan's contemporary history to situate them in a broader historical background.

This research will view democratic transformation in Taiwan as a derivative discourse from Western civilization but one with its own unique continuity in a local historical context. It will also adopt a comparative frame to illustrate the uniqueness of Taiwan's most recent political and social transformation. To this end, Habermas' theoretical framework about the bourgeois public sphere is adopted; more precisely, Habermas theorizes the transformation of the Western bourgeois public sphere as moving from a Literature Public Sphere, through a bourgeois Political Public Sphere to a welfare state formation.[4] My comparative approach will contrast the differences between Habermas' model and Taiwan's historical experiences, and attempt to identify Taiwan's unique historical path.

In a nutshell, this paper will first portray the main cultural schemas of national identities in Taiwan in the Siang-tu wen-siue debates, then situate these schemas in the dramaturgical performance of political contexts at that time, and thereafter delineate the shaping and reshaping of the cultural schemas in the historical discursive formation. Finally, it will use Habermas' comparative frame to shed light on further research topics and on Taiwan's current political development.

Chinese National Identities in the *Siang-tu Wen-siue* Debates

Before we explore the nature of Taiwanese national identities, we should ask a much more general question about what the modern nation is. Most literature regarding Western and third-world countries views nation as a particular modern phenomenon rooted in the French Revolution, and conceived along with the birth of modernity (Feher 1990). While theorizing the modern historical experiences of the Western liberal nation/state, Habermas portrays the bourgeois public sphere as a modern construct that denotes the beginning of a new historical process (1991, 1992).[5] Habermas' public sphere shares a very similar basis with Anderson's national "imaged community" in viewing the communicative media as the most important element of a new social formation. For Habermas, the Political Public Sphere develops after the literature sphere. For Anderson, the foundation of the modern nation is "Print Capitalism." These theoretical lens allow me to situate Taiwan's *Siang-tu wen-siue* Debates in a particular modern historical path in ascertaining the meanings of national identities in Taiwan.

Taiwan's *Siang-tu wen-siue* Debates of 1977–78 is important for exploring the emergence of Taiwanese national identities as well as its relations to others. Since several key figures of the nativist camp left literary pursuits and became directly involved in political protests, this debate became even more important in shaping the subsequent political discourse on the Taiwanese national identity. In the debates, those pro-nativist intellectuals criticized the blind admiration and slavish imitation of Western cultural models (especially the United States), and exhorted their compatriots to show more respect for their cultural heritage as well as greater concern for domestic social issues (Chang 1993: 148). It was the first time since 1949 that pro-nativist writers of the two main ethnic groups, the native Taiwanese (ben-sheng ren) and mainlanders (wuai-sheng ren), joined together to criticize not only modernist writings in the literary sphere, but also the government's modernization project.[6] In the summer of 1977, the country's leading poet, Yu Guang-jhong wrote a short essay titled "Lang Lai le" ("The Wolf is Coming"), openly accusing the nativists of being "leftists."[7] In the age of "the white terror," this fatal charge ignited highly emotional responses from all sides. Consequently, polemic writing about literature and politics began to flood newspapers and literary magazines. This so-called *Siang-tu wen-siue* Debate finally came to an end in the middle of 1978 as a result of threatened government interference.

Although the impact of the *Siang-tu wen-siue* Debates were largely emotional, the participants articulated competing forms of nation building from different perspectives. Those who were supported by the government were promoting a modernization project based upon Western capitalist culture (Peng 1977). Those opposing the official viewpoints were viewing critical realism as an important base, not only for countering the untoward effects of capitalism expansion, but also for forging non-Western national identities (Sie 1979). One pro-nativist thinker, Chen Ying-jhen, regretted that the discussion never ascended a higher theoretical level, never became a "neo-enlightenment" intellectual movement (Chang 1993: 153). More importantly, government intervention in the *Siang-tu wen-siue* Debates signaled the main features of the ongoing Chinese modernization project in Taiwan. Totally detached from China's context, this project was less concerned with China and closer to Western models (the United States in particular) for state building. National ideology, therefore, was simplified along two symbolic lines: "against regionalism" and "against communism" (Peng 1991). Moreover, in the *Siang-tu wen-siue* Debates, Habermas' separation between the literature sphere and the Political Public Sphere is problematic. These debates, in fact, signaled the maturation of a civil society in providing competing viewpoints in the nation-building project. It is evident that the alien authoritarian state exercised its political power in trying to limit these possibilities.

Four cultural schemes on national identities were evident in the Taiwan's 1977–78 *Siang-tu wen-siue* Debates: a Chinese national identity based upon Western liberalism, a Chinese national identity based upon military authoritarianism, a Chinese national identity based upon anti-imperialism, and a Taiwanese identity. As Chatterjee suggests, both thematic and problematic dimensions are evident in the articulations of the aforementioned national identities. However, none of these participants were conscious of the role they were playing. For example, some important writers, such as Huang Chun-ming and Wang Zen-he, whose fiction significantly departed from the modernist fads and depicted rural life with unaffected realism, refused to label their works as nativist (Chang 1993: 151). However, as the debates went on, the differences among the various forms of Chinese national identity and a particular nativist identity, that is, the Taiwanese identity, gradually took shape.

Arguments for Western modernization have been shared by many liberal intellectuals and reform-minded state bureaucrats, and are frequently presented in *The United Daily News* as well as its associated mass media (Ho 1977). This vision of Chinese national identity was represented by literature professor Wang Wen-sing and history professor

Chang Jhong-dong. Wang Wen-sing's articulation represented a positive perception of a thematic part of national identities: only military invention can ruin a country; cultural and political invention do not count as real "invention," as they cannot harm national security (Wang 1978). He viewed Western culture and political models of the liberal tradition as universal for Chinese modernization. Chang Jhong-dong's viewpoint represented concern about its problematic dimension: It is very possible for those nativists to be indulged in regional interests and to ignore the [Chinese] national worldview (Chang 1977). These views had an intellectual genealogy dating from the May Fourth movement in China, within which the liberal form of Western civilization had been applied to reshape Chinese national identity (Cheng 1979). As pointed out by a nativist writer, the liberal reform of Taiwan's "Free China" democratic movement in the 1960s was part of the political practices of the Westernization movement. The national identity resulting from such movements can be seen as a Chinese national identity based upon Western liberalism.

Another discourse on national identities differed from the liberal one in attacking its thematic aspect. This alternative vision was mainly represented in a radical magazine, *Sia-ch'ao* (Chang 1993: 152). A particular form of national identity can be identified in *Sia-ch'ao*'s statements for *Siang-tu wen-siue*. As Yu Tian-chong states, those who are concerned about our social reality and write about it are in fact composing *Siang-tu wen-siue*. Its most important argument is against colonialism, Western flattery, escapism, and regionalism (Yu 1977). Similarly, according to Chen Ying-jhen, Taiwan's *Siang-tu wen-siue* is part of a political, cultural, and social movement toward Chinese nation building (1977). These people supported Chinese unification. In other words, their views on national identities seek to identify Chinese nationalism within the Western critical tradition, Marxism in particular. The thematic dimension of this version of Chinese nationalism is also grounded in Western intellectual discourses supporting socialism and anti-imperialism. During the white terror, they provocatively used taboo terms such as "proletarian literature" (literally, the literature of workers, peasants, and soldiers) and "class consciousness" (Chang 1983: 152). Its problematic dimension is to view the history of the anti-Japanese cultural movement as part of Western imperialism in China (Chen 1977: 66). These Chinese-identified people are strongly against the idea of Taiwan as an independent historical subject with a particular nativist identity. To them a Chinese national identity can be seen as one based upon socialism and anti-imperialism.

Another Chinese national identity was represented by KMT's newspaper, *The Chinese Central Daily* and the military newspaper, *Cing Nian Jhan Shi Bao*.[8] It stressed loyalty to the KMT leadership and emphasized the mission of rebuilding China by overthrowing the standing communist regime. The thematic dimension of such national identities was similar to military authoritarian regimes in many third-world countries, based upon a quasi-Leninist Regime for national building.[9] Its problematic dimension is to emphasize the shameful experiences of KMT's 1949 retreat from mainland China, and to stress its legitimacy in the formal dynastic genealogy of the Han People (Peng 1977). This national identity can be labeled Chinese national identity based upon authoritarianism.

During the period of the Siang-tu wen-siue Debates, Yie Shi-tao (1977), a nativist writer, was the only one who emphasized "Taiwanese consciousness" as an important spiritual core for Taiwanese *Siang-tu wen-siue*. According to Yie, there were two perspectives on such an identity. The first is a realist understanding that "Taiwan was painfully oppressed by foreign colonizers and internal feudal institutions" (Yie 1977: 9). The second is a spirit of resistance based upon "the people's accumulated historical experiences in anti-imperialism and anti-feudalism" (1977: 9). Accordingly, Yie defined Taiwan's *Siang-tu wen-siue* as a realist literature of a national style (1977: 10). He called for all Taiwan's writers to accept critical realism based upon a humanistic concern, to model themselves on the great writers of the nineteenth century such as Balzac, to join the people suffering from the bondage of feudalism and imperialism with a calm and acute realism, and to portray the distress of a nation. "Humanism" and "critical realism" become the thematic dimension of this Taiwanese identity. Its emphasis is on delineating Taiwan's literary history of the past three hundred years as a particular, independent style, different from that of China. It therefore emphasizes that the realism regarding Taiwan should be the basis for literature. On the issue of national identities, the newly emerged Taiwanese identity was related to an "imaged community" but limited to the literature sphere. This is different from the aforementioned Chinese identities that had specific political, economic, and cultural claims.

To summarize, four cultural schemes of national identities can be identified in the period of the *Siang-tu wen-siue* Debates. Each scheme has its own unique logic and may be adopted by various individuals or social institutions. However, these cultural schemes may complement or contradict each other. A person, group, or institution may adopt more than one scheme in the historical context. For example, Peng's collection on the works of the *Siang-tu wen-siue* Debates represents the official positions on national identities, including authoritarianism and liberalism. In Jiang Jie Shi's administration, authoritarianism was supported by the military and police system, led by General Wang Sheng, while Liberalism was supported by the reform bureaucrats, led by Li Huan. Those who supported *Siang-tu wen-siue* possessed cultural schemes including both the anti-imperial Chinese national identity and Taiwanese identity. The possible combinations or interactions of various cultural schemes signify a unique cultural hybridity of Taiwan in the global context.

In the *Siang-tu wen-siue* Debates, the emergence of new visions/schemes may be seen as a challenge to the state's dominant hegemony. However, the state's intervention in the debates signals that the dominant force in the political sphere was capable of intervening in the literature sphere in an attempt to block an emerging hegemony. To ascertain how these schemes have been shaped by and are shaping recent political transformations, we turn our attention to the Political Public Sphere at the beginning of the new historical juncture.

The Dramatic and Performative Fields of the Democratic Movement in the Late 1970s

The year 1978 was a historical turning point for Taiwan. Jiang Jing-guo became the new president and offered a new direction of leadership. In his inaugural address, he stated that "We should set a goal for our common endeavors and gear our concepts and our action to our sense of responsibility for the country [all of China] and the nation and the people, with whom we are joined by ties of blood. . . . We currently need not only to root our economic and national defense construction deeply, but also to work hard for political, social, and cultural advancement" (Guang Hua, ed. 1979: 69). This message signaled a significant shift of the state building project in Taiwan. The new president intended to implement Chinese modernization projects in Taiwan rather than merely treating it as a temporary military base for recovering Mainland China. As for political reform, he stated "we must enlarge political participation, safeguard freedom and human rights and assure that democracy and freedom are based on the will of all the people [the Chinese in both Taiwan and Mainland China] and can be advanced in accordance with moral rationality, dignity of the law, common harmony and sincere solidarity" (1977: 70). Western (especially U.S.) democracy, therefore, became a fundamental part of the modernization project on Taiwan. However, the KMT followed the strategy of the Japanese colonial government in using ethnicity to distinguish between rulers and subjects, the distinction between native Taiwanese and Mainlanders with their specific political and culture history was used to justify the exclusion of native Taiwanese from the government (Li 1987; Huang 1994). Unequal political relations between these two ethnic groups are evident in the practices of defining citizenship. However, the trend of this political shift and the beginning of resistance to political exclusion should be traced to a much earlier time.

In 1977 a new form of political struggle signaled the emergence of civil society and the political manifestation of Taiwanese identities. While it was still illegal to form a political party, hold marches, or stage rallies, a coalition was formed by independent non-KMT politicians known as the *Dang-Wuai*. These politicians, most of whom were native Taiwanese (*ben-sheng ren*), did not in fact have any formal organizational networks and even did not form a common political agenda. *Dang-Wuai* therefore became their common symbol as they called for people to consolidate their dissent toward the KMT in local elections. Their accomplishments in the 1977 local election signaled a maturation of Taiwan's civil society. The *Dang-Wuai* won twenty-two seats in the Taiwan Provincial Assembly and four posts as either county magistrate (county executive) or major in the 1977's election. It was the most significant victory since the KMT regime came to Taiwan after 1949.[10]

According to Foucault (1977), the maturation of modern civil society can be distinguished from its past due to a "modern" style of discursive power formation. As opposed to traditional society wherein power is more systematically exercised by a political ruler as coercive force, in modern society power is shaped interstitially and may either empower or restrain individuals in their local practices. When modern forms of power take shape, the place where the state demonstrates its power may produce subversive consequences and eventually the overthrow of the state. In France, as Foucault explains, the individual body on the guillotine is a case in point when coercive force and resistance meet. In Taiwan, illegal practices of vote collection stations were similar places where forces of state and social riots come together. The state's coercive power was especially evident when local votes were tallied. At that time, the state's agents were trying to produce favorable results for the ruling party. As a consequence, the vote collection station in Taiwan, like the victim's body on the guillotine in early modern France, became the contested place of oppositional powers. In the 1977 Jhong-li Incident, people burned the city's police station and called for a clean election. This incident, as the consequence of new social formation in the existing authoritarian state, signaled that a new form of modern social power had arisen in Taiwan, but, unlike the experience of most Western countries, the discursive modern power did not go far enough to create a revolution (Lin and Chang 1978).

Discourse on liberal democracy became prevalent in the opposition movement after the Jhong-li Incident. Flushed with excitement about their newly discovered social power, the Dang-Wuai activists planned a campaign for the Legislative Yuan election slated for 1978 (Lu 1991). In this campaign, organized political networks became evident, as political activists drafted a common agenda urging civil, political, and social rights. These activists also erected a huge political poster portraying a blue fist and a red slogan, "*Ren-ciuan*" (Human Rights) on a white background. According to the explanation on the poster, these three colors represented "liberty, equality, and compassion" (Lu 1991: 42). The colors of the French Revolution were adopted in Taiwan for political mobilization in the election campaign. Additionally, those colors that were once used by the KMT to distinguish themselves from Communist China were also used by those opposition elites. Western liberal democracy had become an important component of the opposition discourse. The beginning of the liberal discourse can be linked to the "Free China" intellectual movement in 1960s Taiwan, but in the late 1970s this particular intellectual discourse switched gears to mobilize people in Taiwan (Lu 1991; Huang 1994; Li 1987).

However, new events can Change strategic plans, and the termination of formal relations between KMT and the United States is a case in point. Taiwan's 1978 elections were canceled several days before the scheduled vote, due to President Carter's announcement that the United States was severing formal ties with the Republic of China (on Taiwan) in order to recognize the People's Republic of China (Wachman 1994: 139–40). However, the termination of local elections did not suppress the emerging opposition movement. In August 1979, a group of Taiwanese opposition activists organized a journal entitled *Mei-li Dao*. As a way to get around the prohibition against organizing a political party, the *Dang-Wuai* opened a "service office" for *Mei-li Dao Dza-zhi (Formosa Magazine)* that served as a local headquarters for *Dang-Wuai* activists. Rather than passively waiting for the election, these activists began to actively organize rallies for human rights. The newly established local headquarters, therefore, became another contested point between the state's coercive power and the opposition's subversive power.

By establishing formal organizational networks, those opposition activists not only created new opportunities but also experienced new constraints.

At the beginning of 1979, Jiang's administration began strategically to accuse and arrest opposition leaders. Due to this new wave of "white terror" initiated by the ruling party, the new establishment of the *Mei-li Dao* service office became a violent node of confrontation between the state and radical political activists. On December 10, 1979, the Gaosiung Incident—or the *Mei-li Dao* Incident—signaled an outbreak in the confrontation between coercive power and its subversive counterpart.[11] A rally held by the *Mei-li Dao* office in Gaosiung to mark International Human Rights Day, degenerated into a riot in which 183 policemen were injured. The following day, fourteen opposition leaders as well as more than one hundred of the magazine's supporters were arrested. This incident has been generally considered a political trap for opposition leaders since no evidences had been present in the following civil or martial trails (Wachman 1994: 140). Jiang's administration exercised its authoritarian state power to halt the emerging intellectual as well as political movement.

The publicity surrounding the trials represented situational ethnic relations in the Political Public Sphere, as well as political power relations.[12] Two native Taiwanese (*ben-sheng ren*) prosecutors, four native Taiwanese Judges, and a Mainlander (*wuai-sheng ren*) Chief Judge were in charge of this trial. All of the accused and their lawyers were native Taiwanese, who had to counter the arguments from five Mainlander prosecutors and one native Taiwanese prosecutor who had been in charge of this case since the accused were arrested. In fact, only the native Taiwanese prosecutors played the public role of accusing these native Taiwanese elites. The absence of the Mainlander prosecutors combined with the silence of native Taiwanese judges signaled that "the KMT used the native Taiwanese to govern the native Taiwanese but had no confidence in those Taiwanese bureaucrats" (Lu 1991: 316). In this trial, similar to other colonial experiences, the ethnic categories native Taiwanese and Mainlanders became the key taxonomy of KMT's governing, and were also administratively necessary.[13] However, such categories were also the contested sites for indigenous resistance. The resistant spirit of the "Taiwanese consciousness," delineated by Yie Shi-Tao (1979), may be an explanation for the ethnic representation of the defense in the martial trial.

The link between political consciousness and popular resistance was established in response not only to the demonstrations of ethnic power, but also the enactment of state coercion. After the Gaosiung Incident, both the state-controlled mass-media and the educational institutions were mobilized to condemn both the supporters of the *Mei-li Dao*, and their violent behavior. Even before the public trial, a one-dimensional mass-media trial had begun wherein violence and the *Mei-li Dao* were strategically linked.[14] On the first day of the military trial, most family members of the accused lawyer Lin Yi-siung were murdered. It was February 28, 1980, the same date as in 1947 when the KMT military began to suppress an island-wide Taiwanese self-government movement in Taiwan (Kerr 1965). This event has been called the 228 Massacre (Wu 1984). With historical memory interweaving with the rising political terror at that time, this murder shook the Taiwanese public. Right after the end of the court martial, Mainlander (*wuai-sheng ren*) taxi driver Chu burned himself on the front of the Sun Yat-sen Memorial Hall in an attempt to persuade the KMT regime to be tolerant of those reform-minded native Taiwanese political elites (Lu 1991: 352–353). This act of self-sacrifice signaled that the legitimacy of Jiang's regime was challenged, and that Taiwanese identities began to encompass both native Taiwanese and Mainlanders.

The military trial also sparked a new phase of public political discourse on Taiwan's national identities. While the Taiwanese public remained silent, the defending arguments of the accused and their lawyers in the public trial were perceived as heroic at this particular historical moment. Despite being tortured during the interrogation, the accused and their lawyers still clearly stated, "Taiwan's future should be determined by all its eighteen million residents" (Lu 1991: m372; Lei 1987: m270). President Jiang countered their argument by announcing, "We would not allow anybody to use freedom and human rights as an excuse to produce hatred and hamper the security of Taiwan Strait with the intention of suppressing our rights and engaging in the crime of rebelling."[15] These counterarguments indicated two lines of conflicts. Along the line of building a democratic state, the President argued that the Mainlander ruling elites, representing the whole China, had the national mission of laying out the future of the Taiwan. In contrast, the accused and their lawyers argued for a participatory democracy to include all residents of Taiwan (Lu 1991). Moreover, in terms of national identities, all defendants argued for reforms to create a free China with participatory democracy, and denied that they were attempting to promote Taiwan independence. Also, the defendants interpreted their actions as disseminating the idea of self-determination, requesting full political rights for all residents of Taiwan, rather than condoning rebellion and revolts. It was the first time the idea of self-determination appeared in Taiwan's political public discourse. However, it was forcefully condemned due to the political implications of Taiwan independence.

Silence from a suppressed society might normally be viewed as a lack of strong resistance. However, Taiwan's 1980 and 1981 elections revealed that popular sentiments could have unexpected outcomes. In these elections, many family members and defending attorneys of the accused in the court joined the democratic movement (Huang 1994; Li 1989). With their success in the elections of National Assembly representatives and city councilors, it was evident that Taiwanese civil society was able to support alternative visions of state building.

More importantly, a group of young writers and editors continued the efforts of the *Mei-Li Dao* in the elections. They wished to "initiate a mass movement by deliberating knowledge [democratic ideas] and, with the whole power of the society, to create new possibilities in the era of distress" (Lin 1981). The so-called "*Dang-Wuai Cin Sheng Dai*" did not wish to limit the democratic movement to the bourgeois class or political elites (Cheng, ed. 1987). Instead, they wanted ideas of democracy to bring new hope for all people and, with the people's power, to attain the goal of democratic reform.

The New State of the Democratic Movement After the *Mei-li-Dao* Incident

Incidental and unexpected consequences occur frequently in history, as in the case of the *Mei-li Dao* Incident. The democratic movement—once deeply painful—turned out to be a sensitizing event that directed people's attention to political issues, especially the young generation of writers (Chang 1993). Famous prosateur, Lin Shuang-bu, began to write novels detailing human-rights abuses in the 1980s. Novelist Song Jhe-lai began to promote "Literature for Human Rights" (Chen 1988). With the lifting of the ban on magazine publication, political magazines became popular despite the great pressure of governmental censorship (Li 1984). The works of writers from the younger generation enriched these magazines by intertwining and political concerns. In the age of mass media monopolized by the ruling political–military-party system, these political magazines became the important media for the *Dang-Wuai*. They were not afraid to present their ideas, disclose their enthusiasm, or challenge authority.[16] These publications became an important medium for the emerging critical "*Dang-Wuai* Public Sphere."

The Dang-Wuai Public Sphere was situated in a state of authoritarianism wherein the military-police faction of the ruling party was in power (Huang 1994; Li 1987). This powerful political force was represented by General Wang Sheng and his institutional base, the military-police headquarters. At that time, Wang's headquarters had not only produced several domestic and international incidents violating human rights, but had also initiated attacks on the liberal reformers in his own party. At the peak of authoritarianism, the administration reaffirmed that it was inappropriate to establish a new political party, or to have constitutional reform for a representative democracy.[17]

However, the international and social support for "self-determination" (in the sense that Taiwan's future should be determined by all its residents) provided the possibilities for radical democratic reform in Taiwan. Self-determination, therefore, became the most important theme in opposition discourse advocating democracy. However, the strategies used to reach this democratic goal split the opposition elites into two major groups: radicals and reformers (Cheng Fu-shing, ed. 1987). On the one hand, political leader Kang Ling-ciang, representing the reformers, claimed, "I criticize the KMT rather than confront the KMT because, in the current situations, all political reforms have to be accomplished by mediating through the KMT."[18] On the other hand, the *Dang-Wuai Cin Sheng Dai*, the main body of the radicals, did not trust the KMT's sincerity regarding political reform.[19] They wanted to institutionalize the opposition's power and to combine both the election campaign and the mass movement for the purpose of democratic reform.[20] Through criticizing Kang's leadership, the *Dang-Wuai Cin Sheng Dai* established group solidarity and, by using the principle of participatory democracy, created equal opportunities on their own in the struggles with the well-established political elites of the democratic movement (Lin 1983). In 1983, they successfully established a formal political association, the *Dang-Wuai* Writers and Editors Association. In the same year, the failure of the progressive reformers and the success of the radicals consolidated the strength and raised the status of the *Dang-Wuai Cin Sheng Dai* in the democratic movement (Li, 1977; Huang, 1994).

With the rise of the *Dang-Wuai Cin Sheng Dai*, the political radicals began to raise politically sensitive issues in the *Dang-Wuai* Public Sphere. The ethnic question was an important one. Lin Jheng-jie (1981), a Mainlander of the *Dang-Wuai Cin Sheng Dai*, admitted that "the ethnic complex are a historical burden . . . a question to be addressed." He further stated that "the taxonomy of native Taiwanese and Mainlanders should be abandoned. I consider myself a Chinese, also a Taiwanese. Taiwanese should include all native Taiwanese and Mainlander residents" (Lin 1981: 100). At the same time, Chang Deming, a well-established member of the political elite in the opposition movement, expressed the notion that "the Nativist idea of both loving Taiwan and establishing roots in Taiwan is not a narrowly defined ethnic concept to exclude Mainlanders, nor a denial that the ancestors of the Taiwanese are also from China" (Chang 1981). Similar to "Taiwanese consciousness" in the literature sphere, Taiwanese identity is mainly land-based. Such identities became an important vision for the democratic movement to use in overcoming ethnic barriers.

However, these considerations were absent from the speeches or statements of President Jiang Jing-guo. In 1976, Jiang denied the legitimacy of the term "Taiwanese" (Li 1987). In an interview in 1978, he denied any ethnic preference in officer appointments and claimed that "all people in Taiwan are Chinese."[21] In practice, although he continued to use native Taiwanese elites to relieve ethnic tensions, he allowed the Mainlander elites to play major roles in the Central Government (Huang 1994). In the late 1970s and early 1980s, Jiang's approach to Chinese national identity viewed "Taiwanese" and "native Taiwanese ben-sheng ren" as non-existent and illegal terms and emphasized the legitimacy of his regime in representing all of China.

These policies of allowing only the terms "China" and "Chinese" and delegitimizing the terms "Taiwan" and "Taiwanese" also captured the attention of the *Dang-Wuai Cin Sheng Dai*. As Lin Shi-yu, a key member of the *Dang-Wuai Cin Sheng Dai* observed, "In various official statements or talks, the KMT administration usually refuses to use the word 'Taiwan.' They always refer to Taiwan as 'a (military) base for recovery.' In the eyes of the administration, these four words become the whole meaning of Taiwan" (1982: 5). He raised further questions: " 'A base for recovery' is for 'recovering mainland and unifying China.' In addition to this, why don't the industrious efforts and stupendous achievement of the Taiwan's people as well as the society they have established have any meanings for themselves?" (1982: 5).

In the early 1980s, the *Dang-Wuai Cin Sheng Dai* established their position in the democratic movement, forged their visions, and produced critical reflections on political reality. The formation of a *Dang-Wuai* Public Sphere served as a communicative medium through which to shape collective identities. Through this process, they were able to follow the principle of participatory democracy in forming political organizations. Engaging in political debates with the well-established political elites in the democratic movement, opened a door for them to formulate internal consensus. While applying the principle of participatory democracy when establishing the *Dang-Wuai* Editors and Writers Association, they were able to transform individual thoughts into collective action. At the same time, they began to challenge the administration's policy on state building and national identities.

The Debates on Taiwanese Identities and Chinese Identities

Some seemingly incidental events may prompt people to critically reconsider situations that have long been taken for granted. The 1983–1984 debates on Taiwanese and Chinese identities in the *Dang-Wuai* Public Sphere are a case in point.[22] These debates were initiated by Chen Ying-zeng's comments on the actions of the famous folk singer, Hou De-Jian. Hou De-Jian, the composer of the popular song "The Descent of the Dragon" (a symbolic term for Chinese ancestry), visited Mainland China (Chen 1984). According to Chen Ying-zeng, Hou was "going to see an ancestral country, whose essence has long been running in his blood, what has mused him in his dreams and provoked him by its tides and waves, and which has been formed by thousands years' history and culture." In the same essay, Chen regarded " 'Taiwan and Taiwanese' [national]ism" as an error, attacked such consciousness as "prevailing among a small group of bourgeois intellectuals" and criticized it for excluding Mainlanders from the democratic movement (1986). Chen's and subsequent articles supporting Chinese national identity strongly attacked "Tai-wan Min Ben Jhu Yi," an issue raised by Liao Wen-yi, a Taiwan independence activist in postwar Japan.[23] Moreover, Chen's essay implicitly echoed the observation of Lin Chen-jie, one of the *Dang-Wuai Cin Sheng Dai*, about *Dang-Wuai*'s old, middle, and new generations, differing opinions regarding Mainlanders.[24] However, new visions of Taiwanese consciousness were presented in the debates. For example:

> We will blast the myth of [Chinese] nationalism,
> let people learn to place themselves in a position of well being,
> we will let all Chinese people realize
> things are not necessarily to follow the old rule,
> things are Changeable,
> as long as they can become better. (Lin Jhuo-shui 1984: 239)

This poet, with strong humanistic concerns but critical of Chinese nationalism, represents the reflections of the *Dang-Wuai Cin Sheng Dai* toward Chinese national identities, which emphasize thousands of years of the Han people's cultural tradition. In the Taiwanese identities debates, such reflections were based upon criticisms of Chinese national identities. These criticisms focused on the following dimensions of Chinese national identities: first, its ethnocentrism; second, its propensity for invasions and exclusions; third, its insults to the minorities during its historical formation; fourth, its strong racial discrimination; fifth, its factual and moral bases, which critics saw as unfounded; and finally, its fictitious history of racial purity.[25] In general, these criticisms bemoaned the unfairness of underestimating Taiwan's history and culture as well as a growing unwillingness to support Chinese nationalism.

Additionally, critical questions about Taiwanese consciousness pushed independence writers and the *Dang-Wuai Cin Sheng Dai* to reconsider their meanings. Chen Fang-ming, a famous Taiwanese poet and historian, restated Yie Shi-tao's interpretation of Taiwanese consciousness and illuminated the thematic aspect of national identity, rationalism, subjectivity, and progress. As Chen stated, "The nearly four-hundred-year-old Taiwanese society has been shaped by interactions between the diligent farming of immigrants and the frantic exploitation of colonizers . . . Although those rootless colonizers came and went, those rooted immigrants, under particular socioeconomic conditions, and with their subjective willingness to struggle, finally forged a solid nativist consciousness. It becomes what we now call Taiwanese consciousness" (1985). It was the first time an independent historical vision of the Taiwanese people had been articulated. Chen's reinterpretation signaled an outgrowth from the *Siang-tu wen-siue* Debates, through which a socialist concern about the people and land formed the basis for defining the Taiwanese identities.

The Taiwanese identity debates further enriched the various perspectives of Taiwanese national identities and presented a dynamic, reflexive picture of those identities. These ideas include: seeing both the Taiwanese and the Chinese nations in relative terms; juxtaposing both Taiwan's and China's histories and cultures; viewing Taiwanese identities as dynamic (i.e., the identities that does not exist today may develop another day); seeing Taiwanese identities as a foundation of the democratic movement; and claiming such identities as a prospective concept, based upon Taiwan's subjectivity and historicity in recounting Taiwan's international relations.[26] To summarize, Taiwanese consciousness based upon the land and the willingness of its residents was articulated as relative and equal to Chinese consciousness. Such a consciousness, as the basis for Taiwanese national identities rejected the assumption that

Taiwan's historical possibilities depended on China, and expected to place Taiwan's future in the hands of its own people (Cheng, ed. 1987). Consequently, both the principle of "self-determination" and the emphasis on social democracy both became the important contents of Taiwanese national identities in the *Dang-Wuai* Public Sphere.

The emergence of Taiwanese national identities reformed power relations within the *Dang-Wuai*. Those Chinese-identified people, most of them affiliated with *Cia-chao*, began to be confronted by the *Dang-Wuai Cin Sheng Dai*. At the same time, those supporting the social democracy-based Taiwanese national identity formed a new political association, the *Sin Ch'ao Liu*. They cooperated with the independence-minded liberal radicals, and together formed the main body of *Dang-Wuai* Editors and Writers Association (TWEWA). Generally speaking, the TWEWA became a united front for the *Dang-Wuai Cin Sheng Dai* to emphasize Taiwanese identities, insist on self-determination, mobilize people, organize a new political party for the democratic movement, and risk direct conflicts with the KMT (Li 1987).

Several contingent events helped speed Jiang's political reform. Internationally, in 1984 writer Henry Liu was murdered in collusion with Taiwan's military intelligence organs, possibly under the direct influence of Jiang Jing-guo's supposedly ambitious son, Jiang Hsiao-wu (Wachman 1991: 142). The U.S. government condemned the action, and the U.S. Congress passed legislation urging Taiwan to practice participatory democracy. Domestically, a large economic scandal broke out due to corruption among military personnel, state bureaucrats, and party workers. Moreover, new social movements began to emerge to promote various social and economic issues.[27] Even worse for Jiang's administration, Premier Shun, who had established a very good reputation in society, suffered a stroke in 1985. In this critical time period, Jiang decided to speed up the path of democratization, and led a "reformation from above" in his party, the KMT, as well as the government.[28]

Jiang's reforms reshaped the governing strategies of his administration. He first suppressed the power of the military-police and put civilians in charge of the Ministry of Defense (Huang 1994). The power of Taiwanese KMT members in decision-making was also significantly increased.[29] Moreover, in his interview with a visiting group from *The Washington Post* in 1986, Jiang indicated important shifts in his state building project. For facilitating democratization, he stated "we will carefully and actively study issues relating to the formation of new political parties and the termination of Jhieh-yen" (1986: 27). About his successor, he stressed that "The Republic of China is administered pursuant to the Constitution and the next president will be elected accordingly. The members of my family will not run for the next presidency" (1986: 30). As for the blueprint of the state-building project, he claimed that "The Republic of China and the United States have similar political systems" (1986: 30). A participatory democratic reform became evident in Jiang's interview.

The new democratic reform was accompanied by a new shift in dominant national identities. By the end of 1986, while Jiang was meeting with twelve Taiwanese elder elites, Jiang told them, "I have lived in Taiwan for more than forty years; in consequence, I am also a Taiwanese" (Li 1986). As Cline pointed out "It did not mean he was not Chinese, because the Taiwanese are also Chinese. They also would like to see all of China reunited on principles that they have put in practice in the small island of Taiwan" (Cline 1989: 138). However, it is evident that Jiang adopted a land-based Taiwanese identity. And while acclaiming Taiwanese status, he also laid out a unified future with China for the people in Taiwan. These claims indicated an important shift in dominant national identity. President Jiang apparently accepted the 1982 claim of the imprisoned *Mei-Li Dao* elites: "To practice democracy in Taiwan is far more important and necessary than to create a chance for a united China."[30] For President Jiang, a land-based Taiwanese identity did not exclude blood-based Chinese identities. He decided to take on the Taiwanese status to engage in democratic reform, which legitimized his "Chinese" administration.

In contrast to the reform of the KMT, the power rearrangement *in Dang-Wuai* represented a reformation from below. It was especially evident after 1984. Any attempt of the progressive reformers to negotiate with the KMT was criticized by the *Dang-Wuai Cin Sheng Dai*; these attempts were even discouraged by their mobilization of mass demonstrations.[31] After encountering these challenges, the leaders of the proprogressive reformers also adopted radical strategies for forming a new political party in order to subdue the criticisms and pursue a consensus with the *Dang-Wuai Cin Sheng Dai*. Such power dynamics occurred in the *Dang-Wuai* Public Policy Association (a united front of *Dang-Wuai* legislators including radicals and reformers), in the negotiation with KMT delegates, and in public speeches explaining the importance of a new political party. This power dynamic, along with pressure from overseas Taiwanese associations and the support of the U.S. Congress, produced a moment ripe for establishing a new political party. In 1986, the new political party, the Democratic Progressive Party (DPP), was established during a *Dang-Wuai* meeting.

The 1986 election, signaled a major shift in state building and nation formation in Taiwan. In terms of state building, this election, similar in style to the American two-party democracy, resulted from the competition of two political tendencies: President Jiang's reform from above in the KMT, and *Dang-Wuai Cin Sheng Dai's* reform from below in the *Dang-Wuai*. The initiators of these trends were the paternalistic leaders of the existing authoritarian regime and a new generation of critical intellectuals devoted to social democracy. In terms of national identity, President Jiang apparently favored Chinese national identities based on liberalism. Additionally, his new meaning of national identities also included a realistic recognition of Taiwanese national identities based on land. It is also important to note that he was considering the future of Taiwan under the frame of a united China. More precisely, for President Jiang, a Chinese national identity was the guiding frame for any Taiwanese national identity. However, the *Dang-Wuai Cin Sheng Dai*, and *the Sin Ch'ao Liu* in particular, considered the two kinds of national identities to be equal and wished to see an independent future for Taiwan. The differences in defining the relations of these two national identities became the focus of the debates in the Political Public Sphere after the 1987 lifting of martial law.

Competing Notions of Taiwanese National Identity Since the 1987 Sie Li-fa Debate

The debates in the literature sphere, as well as in the *Dang-Wuai* Pubic Sphere, continued to have an impact on the Political Public Sphere in forging Taiwanese national identities. Writers' reflections on Taiwan's history and future, frequently enriched Taiwanese national identities and gave it new meanings. During the 1987 Sie Li-fa Debate, the main political forces in Taiwan had to reconsider the relationships between Taiwanese and Chinese national identities, as well as the meanings of respective national identities.

The 1987 debates focused on Taiwanese painter Sie Li-fa article, "The Blindness of Taiwanese Intellectuals: An Analysis from the Perspective of the 228 Event" ("*Chong Er-Er-Ba Shi-jian Kan Tai-wan Jhi-shi-fan-zi De Li-shi mang-dian*").[32] This essay stated that "Since the Ming Dynasty, the relations between Taiwan and China have been situated in two positions: dependence and confrontation. Dependence frequently resulted in the injury and humiliation of Taiwan; confrontation will produce failure as well as surrender . . . . Recently . . . the success of Taiwan's independent economy . . . encouraged a thorough reflection on the political and cultural domains. People will finally find that, to avoid dependence on and confrontation with China, we have to wipe out Chinese sentiment, terminate all old relations [with China], and follow a path of the economic development to move toward independence and autonomy" (1988: 93–94). By urging an open attitude toward Western liberal societies, Sie made a much clearer distinction between Taiwan and China and called for an independent Taiwan.

In the same essay, Sie strongly criticized Taiwan's democratic movement. According to Sie "For many years, the Taiwanese merely adopted a political approach to address the issues of cultural identities. As a result, the democratic movement was limited to election campaigns and became a pure election movement. Along this line, the democratic movement has accomplished nothing except continuously following a political path of dependency on the existing political regime as well as its cultural identities" (1988: 88–89). He urged Taiwanese intellectuals to critically reconsider their own situation, to support the people, and to free themselves from dependency on Chinese nationalism to help promote independent Taiwan (1988: 94). Sie expanded the scope of Taiwanese national identities to include political, economic, social, and cultural dimensions, and to escape from the Chinese nation/state building in the quest for an independent Taiwan.

The debates detailed the position of the Sin Ch'ao Liu, one major faction in the DPP, on Taiwanese national identities. At that time, Shieh's article was strongly criticized by many writers with Chinese identities, especially by those affiliated with the democratic movement because they believed that aligning oneself with Chinese national identities was important for anti-imperialism. To counter such criticisms, Lin Jhuo-shui, the General Editor of DPP's magazine and a key member of the Sin Ch'ao Liu, defended Sie's position. According to Lin, Sie's view of an "independent Taiwan" is similar to the idea of "self-determination" in the DPP's constitution and may not necessarily contradict the viewpoint of those pursuing a united China. Also, Taiwan's experiences in "anti-feudalism" and "anti-imperialism" cannot be separated from its history, especially in the Japanese colonial period. By the same token, Lin approved of Sie's viewpoint on using Western concepts to empower people in the struggles of anti-imperialism and anti-feudalism. Furthermore Lin sided with Shieh's position and urged Taiwanese intellectuals to "stand with all Taiwanese people to see the state as the means and the people as the end [in the democratic movement]" (1987). Through Lin's arguments, it is evident that pro-Taiwanese national identities statements in the 1987 debates grew out of the main ideas of

Taiwanese identities in the 1977–78 *Siang-tu wen-siue* Debates. Conceivably, Sie and Lin are more open to Western liberal ideas than most anti-imperialist nativist writers in the 1977–78 debates.

In addition, Lin's defense of Sie's essay sheds a new light on Taiwanese national identities, especially regarding the relationships among people, intellectuals, and the state. More precisely, it pointed out new questions: What does Lin mean by the state as the means and the people as the end? How do intellectuals, particularly political and social activists, play a part in the means-and-end relationship? In a newspaper article, "Lin Jhuo-shui Wen Ji" (1991), Lin provided a much clearer explanation of his ideas (1992). He first portrayed Taiwan's society as "restlessly energetic, frequently changing. . . [It] represented a state of disorder and uncertainty after the lifting of martial law." Political participants, according to Lin, "should play a role as mediators to unite those various conflicting social forces . . . to respect each other's individuality, to effectively and sufficiently pay attention to the particularity of each changing moment." As for the relations between society and the state, he stated that "the constitution is the negotiated result of various conflicting social forces and, in consequence, its rules should not be rigid but must be frequently adjusted in keeping with the new social formation in time and place" (1992: 180). Generally speaking, Lin portrayed Taiwan's society as a pluralistic and uncertain one, and, accordingly, suggested a reform from below, with a people-oriented rather than elite-oriented approach to state building. More precisely, Lin's view on the relationship between intellectuals and the people is close to Gramsci's idea of organic intellectuals, speaking for the oppressed class or people.[33] While the Sin Ch'ao Liu participated in an exhibition on performing arts, Lin Jhuo-shui used the concept of "Total War." This means empowering the people, who had been marginalized and oppressed by the state, to engage in a political, social, and cultural confrontation with the ruling authority, and to attain their historical and social agency (Lin 1989). In agreement with *Sie Li-fa*, the Taiwanese national identity specified by Lin Jhuo-shui is one based upon "social democracy" (Li 1991).

However, Lin's viewpoints on national identities faced new challenges within his own party in the late 1980s. To detail the power relations of these competing visions, we have to return to an earlier time. Since the early 1980s, *Dang-Wuai Cin Sheng Dai*, and the *Sin Ch'ao Liu* in particular, successfully led reform from below in the democratic movement, and controlled the power of the newly established DPP party (Huang 1992). They did not face significant challenges until the imprisoned *Mei-li Dao* leaders, Change Jhu hong and Huang Sin-jie, were released. These two formed a new faction, Pan Mei Li Dao, within the DPP, to counter the power of the *Sin Ch'ao Liu*. Moreover, Chang (1989) offered a new vision for state building based upon liberalism and progressive reform. More precisely, Chang's strategies were to focus upon local election campaigns. In his view, victory in local elections constituted an important step toward accumulating the power base for accomplishing self-government in local counties. By winning victories in most local counties, the DPP would be able to push the central government to accept self-determination, and finally be granted independence. Chang strongly argued against the *Sin Ch'ao Liu* socialist viewpoint and its insistence upon an independent Taiwan. It is conceivable that Chang may not have given up the vision of Taiwan as an independent sovereignty. However, he did not openly address concerns about national identities, which were possibly related to the complicated problem of varying cultural and historical memories among various ethnic groups. Consequently, Chang attempted to use liberal-based state identities to challenge arguments for Taiwanese national identities. A bourgeois state-building strategy was supported by *Pan Mei-li Dao* to challenge the *Sin Ch'ao Liu*'s socialist nation-building strategies.

The new challenge from *Pan Mei-li Dao* pushed the *Sin Ch'ao Liu* and its allies to defend their position on Taiwanese national identities in the Political Public Sphere. In 1989, *Sin Ch'ao Liu* and other independence-minded groups formed a united front, the New State Alliance, for the upcoming election campaign. They laid out a state-building project in a draft Constitution of the Republic of Taiwan, with the goal of establishing an "Oriental Switzerland" on the basis of self-determination (Hsu 1990). Since then, Taiwanese national identities in the Political Public Sphere have represented an alliance of not only domestic organizations, including the *Sin Ch'ao Liu* and Association of Political Prisoners, but also various overseas Taiwanese associations.[34] Alliance members viewed the election campaign as an opportunity to promote their ideas. With this newly established group in the 1989 election campaign, Taiwanese national identities became a sufficiently competitive discourse to challenge the dominant status of Chinese national identities in the Political Public Sphere. A new state-building project was also established without dependence on or confrontation with China.

After the death of Jiang Jing-guo, Li Deng-hui became the president of the ROC in 1988. As the first Taiwanese president, he had the Taiwanese intellectual mandate to lead them to freedom from political despotism. A popular strategy had been to gain backing of the ethnic core to support a particular political leader in Taiwan. As a consequence, the state and the ruling party became contested fields in the ongoing political reform. With the situational support from reform-minded state bureaucrats, university students, intellectuals, the opposition party, and the general public, President Li fought and overcame various countering forces within the KMT and the state (Chou 1993; Li 1992). Moreover, benefiting from popular support, he took strong action to change/reconfigure ethnic power relationships within the state and his party. In Li's administration, native Taiwanese became dominant and Mainlanders became adjunctive. However, under Li's rule, the main tensions in the Political Public Sphere no longer involved ethnic distinctions, but centered on national identities.[35]

The above political changes reshaped visions of national identities and their corresponding power relations in the Political Public Sphere.[36] The New Party was formed by Chinese-identified ex-KMT members, and the "non-mainstream" faction of the KMT. These groups shared similar identities with the New Party. However, they also, represented the ideas of Jiang Jing-guo in the early 1980s, which reflected ambivalence about authoritarian Chinese and liberal Chinese identities. According to William's model of hegemony, Chinese national identities can be seen as residue. President Li and his followers represented a dominant hegemony reflecting the ambivalence between liberal Chinese national identities and Taiwanese national identities. The DPP stood for an emerging cultural hegemony with a particular ambivalence developing between liberal state identities and socialist Taiwanese national identities. Within the various national identities, democratic state-building and liberalism become the denominator for these political networks.

However, during this time period, Taiwanese national identities risked excluding most people with entirely Chinese identities. The consolidation of Li's administration signaled a hegemonic shift in Taiwan. A new set of power relations underlying the cultural schemas of national identities was also evident. Those people with exclusively Chinese national identities separated from KMT and joined the newly established New Party. They represented the old KMT hegemony, with its ambivalence between the authoritarian and liberal Chinese national identities. As for Li's administration, an ambivalence between liberal Chinese national identities and Taiwanese national identities was evident. The DPP, the largest opposition party, displayed ambivalence between liberal state identities and the Taiwanese national identities based upon social democracy. A new cultural hybrid of national identities became evident, of which the nonexclusive Taiwanese and liberal

Chinese national identities were the prevailing hegemony. This new form of Taiwanese national identities, independent of Chinese national identities, can be seen as an emerging counter-hegemony.

It should be noted that although the three main political parties represented different national identities, these identities are not exclusive. In fact, these parties are represented as a coalition of people with various political and social concerns. National identities were merely one of these concerns. Democracy was both the medium and the outcome of the transformation of national identities in Taiwan. As a result, Jiang Jing-guo's 1970 interpretation of Chinese national identities lost ground, *Mei-li Dao*'s version of Chinese national identities were dominant in the early 1980s, and the *Sin Ch'ao Liu*'s version of Taiwanese national identities was emerging in the late 1980s. The radical visions have significantly shaped Taiwan's politics, and are very likely to continue to play a role in the country's future.

During this time period, Taiwanese national identities became the basis for mutual trust between the opposition elites in the democratic movement and the KMT reformers led by Li Deng-hui. This trust limited the possibilities of radicals and created a situation that Wachman has labeled progressive reform. However, democracy means different things to different political parties. The DPP had experienced reformation from below and earned support through mass mobilization. It was much more comfortable with the idea of participatory democracy and had a tendency to support welfare policies that addressed the needs of "all people" in society. The KMT, having experienced a reform from above, presented a style of elite-oriented democracy, a new form of Chinese paternalistic rule. The different historical experiences of both parties provided an important base to displace the ideas of Western democracy with their own institutional narratives.

Conclusion and Present Concerns

This research has recognized the epistemological limitations of Western theory, and did not limit the historical analysis of a non-Western country by Western theory. Rather, it adopted Chatterjee's analytical angles on nationalism, the thematic and problematic, and Habermas' theoretical frame on Western democratization, the transformation of the bourgeois public sphere. This research, situated both analytic angles and the comparative frame into Taiwan's historical context, and theorized Taiwanese national identities according to the analysis of Taiwan's intellectual narratives. Further, it constructed an independent subjectivity/historicity of a non-Western country with its own particular intellectual genealogies as well as historical path.

In retrospect, it is evident that the political transformation in Taiwan can be seen as a consequence of competition between old and new ideas. However, the new ideas were rooted in the existing ones. New visions of Taiwanese national identities, included liberal democracy and Taiwanese identities, both being important foundations of the democratic movement. In the early 1980s, the *Dang-Wuai Cin Sheng Dai* wished to "follow the Predecessors steps and succeed the democratic tradition." As a result, they incorporated energetic social forces into the existing intellectual democratic movement, which had its roots in the "Free China Reform" (Deng 1981). Within the interactions between the political activists and nativist writers, a collective sense of Taiwanese identities was forged and, through critical reflections on Taiwan's history and society, created a new vision for Taiwanese national identities. This vision, situated in a particular international context and granted support from both overseas and domestic critical intellectuals, contributed to an irresistible momentum for Taiwan's democratization. During this time period, Jiang Jing-guo also joined the movement of being Taiwanese. As a result, after 1986 the momentum of this new vision ran like a "huge river," helping to establish Taiwanese identities and create a democratic Taiwan.

In the 1980s, a unique path of Taiwanese national identities emerged. In the beginning, it was first forged in the tradition of humanistic concern and critical realism in the tradition of the *Siang-tu wen-siue.* It appeared as Taiwanese identities in the Literature Public Sphere. Around the time of the *Mei-li Dao* Incident, activists in the democratic movement began questioning certain political and cultural practices. These included unequal power relations based upon ethnic taxonomies, and nation/state-building project based only upon the ruler's historical experiences. Since then, belief in democracy and concern for oppressed peoples have become the basis for formulating political criticism and reshaping the national imagination. While reconsidering the experiences of oppression, the opposition moved beyond the limitations of ethnic categories. After critical reflections on Chinese national identities, they tried many new directions: to free themselves from ancestral destiny, to enrich Taiwanese national identities with ideas for social democracy, to pursue a better future, to lay out a state-building project for people in Taiwan, to avoid dependence or confrontation with China, and, finally, to pursue an independent future for Taiwan.

With the progress of democratic reform, the newly emergent Taiwanese national identities also challenged the dominant status of

Chinese national identities. In this process, the dominant hegemony of the ruler changed. More precisely, the establishment of democratic reform and Taiwan-based state-building projects helped mold a national consensus between the ruler and oppositional elites. However, the rulers and Taiwan nationalists had conflicts over Taiwan's future and its relations with China. As a result, when Taiwan-based state-building orientation became a consensus, a state identity based upon liberalism replaced the original Chinese national identities and became the dominant cultural hegemony in the Political Public Sphere.

When we return to the current concern about Li's compromise with pro-independence forces and his suppression of underground mass media, the findings of this research suggest that Li's actions are consistent with his identities. With liberalism being part of the newly established dominant hegemony and the rise of bourgeois elites in the dominant political parties, the new conflicts emerged along class lines rather than national identity lines. To cope with social movements on welfare issues, Taiwan's democratization has developed basic institutions similar to those in the Western democracy, and is being pushed in the direction of becoming a welfare state. Along this line, Li's administration began to work on the project of national health insurance in the late 1980s, and implemented this project in March 1995.

After a period of Taiwanization and democratization of the KMT regime, people in Taiwan find themselves situated in a hybrid "post-colonial" condition of Chinese feudalism and Western imperialism, especially in a sense of historical continuity. The new situations may remind us of the claim of Yie Shi-tao in the "Introduction to the History of the *Siang-tu wen-siue*": "With calm and critical realism, [intellectuals or writers should] side with people bounded by the chains of feudalism and colonialism . . . to understand the distress of a people" (Yie 1979: 10). This vision has enriched Taiwanese national identities of the 1980s and has created new possibilities for the democratic movement. It is conceivable that ideas of Taiwanese identities will remain important in the continuing effort to enrich culture and to pursue an independent Taiwan, along the lines of its historical experiences with "anti-imperialism" and "anti-feudalism" in the present post-colonial context.

Notes

1. Wachman argues that this developmental path can be perfectly explained by Huntington's five phases of the "third weave" in democratic transformation.
2. One of the best examples is Rafael's research on translation and Christian conversion in Tagalog society under Early Spanish rule (Vincente 1993).

3. For a detailed account of the contrast-oriented approach, please see Reinhard Bendix (1964); Theda Skocpol and Margaret Somers (1980), pp. 174–197; and Charles Tilly (1984).
4. For the references about the public sphere, please see Jurgen Habermas (1992); Jurgen Habermas (1991); and Calhoun, Crag ed. (1992).
5. Although many contemporary western scholars such as Eley, Fraser, Calhoun, and Benhabib point out the exclusivity and multiplicity of the Public Sphere, one very important missing ingredient in Habermas' theory of Public Sphere is the idea of nationalism (Calhoun 1990).
6. In this paper, native Taiwanese, "pen-sheng ren" represents the ethnic group of Taiwan's nativists, who have been in Taiwan or have direct families tie with people in Taiwan during and before the Japanese colonial period. Mainlanders, "wuai-sheng ren" are the newcomers who bacame Taiwan's residents after Taiwan was taken over by China after World War II.
7. For the references of the *Siang-tu wen-siue* Debates, please see Peng (1991); Hou (1977); and Chang (1993).
8. Peng Ping-kung ed. (1977): this collection was the most important one to affirm the official standpoints in the period of the *Siang-tu wen-siue* Debates.
9. For specific discussion on the form of political regime in Taiwan, please see Cheng (1989); for discussion on authoritarian regime, please see Ruo Lin Jhen Chang (1992).
10. For this period of Taiwan's history, please see Lu (1991); Lai ed. (1987); Li (1987); and Huang (1994).
11. For the references on Mei-li Dao Incident, please see Lu (1991); Lai (1989); Huang (1994); and Li (1987).
12. For the references on the trials, please see Lu (1991) and Lai (1989).
13. About the ideas of administratively necessary taxonomy, please see Cohen (1987), pp. 224–253; Rafael (1993), pp. 185–218; and Fasseur (1994), pp. 31–56.
14. Their arguments were evidenced in almost every mass media in Taiwan at that period of time. However, a more descriptive account can be seen in Lei (1987), pp. 187–191.
15. President Jiang's speech on April 3 (1980). This quotation comes from Lei (1987), p. 189.
16. For details about the *Dang-Wuai Cin Sheng Dai*, please see Deng (1981); Lin (1981), pp. 104–133; and Sie (1983).
17. For details, please see *Independent Evening News*, 1982, June 14, September 25, October 15, and November 24.
18. Interview with Kang Ling-ciang (1980), *Independence Evening Post*, December 5.
19. Interview with Chu Yi-ren (1983), *Independence Evening Post*, July 6.
20. Interview with Lin Shi-yu and Chu Yi-ren (1983), *Independence Evening Post*, July 6.
21. In a dialogue with David Lee, Editor of Readers Digest (1979), in Kwang Hwa, "The Republic of China is on the Move," Taipei: Kwang Hwa Publishing Co., p. 14.
22. For references about these debates, please see: Chen (1985) and Kuo (1984).

23. For references about "Tai-wan Min Ben Zhu Yi" please see Wu (1994) and Huang (1994), pp. 195–236.
24. According to Lin Jhen-jie's observation, the activists of the old generation usually do not believe that Mainlanders will join the Democratic movement. Those of the middle generation see the joining of Mainlanders as unusual. Those of the new generation do not have the conflicts between native Taiwanese and Mainlanders. Please see Lin (1981), p. 126.
25. For the details, please see Lin (1984) and Chen (1985).
26. These arguments are summarized from the following readings: Lin (1984); Liang (1984); Tsai (1984); Yie (1983); Cen (1984); and Kuo (1984).
27. Michael Hsiao ed. (1989).
28. About the period of political Change, please see Li (1987); Wachman (1994); and Huang (1994).
29. Taiwanese vice-president Lee became his formal representative for managing domestic and international affairs. Moreover, Jiang took a huge step toward reorganizing the power structure of the state and KMT in 1986. More Taiwanese elites were included in the decision-making. In 1986, 45% of KMT's central committee members were Taiwanese. For details, please see Huang ed. (1987).
30. *Independence Evening Post,* September 29.
31. For references to the following analyses, please see Huang (1994); Wachman (1994); Li (1989); and Huang (1992).
32. For a delineation of this incident, please see Cheng (1988).
33. Gramsci specifically referred to the organic intellectual for the working class. However, in the tradition of English cultural studies and its American counterpart, the term organic intellectual has been widely applied to the intellectuals for the oppressed people. Here, this paper adopts the broad definition. For the original definition, please see Gramsci (1971).
34. For references on overseas Taiwanese independence movement, please see Chen (1992) and Li Fung-Tsun et al., *Fun Chi Yun Iung.*
35. The following analysis is based upon Chou (1993); Huang (1994); and Li (1987).
36. The following analysis is based upon: Huang (1994); Wachman (1994); Lee (1994); Ho (1993); and Chou (1992).

References

Anderson, Benedict R. O'G (1990). "Old State, New Society: Indonesia's New Order in Comparative Historical Perspective," in *Language and Power: Exploring Political Culture in Indonesia.* Ithaca and London: Cornell University Press.

——(1991). *Imaged Communities: Reflections on the Origin and Spread of Nationalism.* London and New York: Verso.

Bendix, Reinhard (1964). *Nation-Building and Citizenship: Studies of Our Changing Social Order.* New York: Wiley.

Bhabha, Homi K. ed. (1990). *Nation and Narration*. London and New York: Routledge.

Boon, James (1977). "Bali-tje: A Discursive History of the Early Ethnology (Post-1597)" in *The Anthropolitical Romance of Bali, 1597–1972*. Cambridge: Cambridge University Press, pp. 10–34.

Chang Jhong-dong (1977). "Siang-tu Ming-zu Zi-li-zi-ciang," in *Jhong Guo Lun Tan*, vol. 5, no. 2.

Chang, Jhu-hong (1989). *Dao Jhi Jheng Jhi Lu: You Di Fang Bao Wei Jhong Yang De Li Lun Yu Shi Jian*. Taipei: Nan Fang exception.

Chang De-ming. *Min Zhu Tuang Jie Ai Tai Wan, Ling Wei Dang-Wuai*. Taipei: Ling Jheng-Jie.

Chang, Wen-jhi (1993). *Dang Dai Tai Wan Wen-ciue De Tai Wan Yi Shi*. Taipei: Independent Evening News Publishers.

Chang, Yvonne Sung-sheng (1993). *Modernism and the Nativist Resistance: Contemporary Chinese Fiction from Taiwan*. Durham and London: Duke University Press.

Chatterjee, Partha (1986). *Nationalist Thought and the Colonial World*. Minneapolis: University of Minnesota Press.

Chen Ying-jhen (using the pen name Siu Nan-chung) (1977). "Siang-tu wen-siue De Mang Dian," in *Taiwanese Literature*, new edition, vol. 2.

——(1979). "Wen-ciue Lai Zhi She Hui Fan Ying She Hui in *Sian Ren Jhang Siue Ji*. Taipei: Sian Ren Jhang.

——(1984) "Siang Zhe Geng Kuan Guang De Shi Ye," in *The Progressive Weekly*, vol. 12.

Chen Fang-ming (1984). "Sian Jie Duan Tai Wan Wen-siue Ben Tu Hua de Wen Ti," in *Taiwanese Literature*, vol. 86.

——(1985). *Tai Wan Yi Shi Lun Jhang Siuan Ji*, LA:

——(1988). *Mei Li Dao Shi Jian Yu Tai Wan Wen-siue Hang Siang*. Gaosing: Dung Li.

——(1988). "Huang Wu Yu Fong Riao," in *Sie, Li-fa Chong Shu Tai Wan De Sin Ling*. Taipei: Freedom.

Chen Ming-cheng (1992). *Hai Wai Tai Du Yung Dong Si Shi Nian*. Taipei: Independence Evening Post Publisher.

Chen Shu-hong "Tai Wan Yi Shi-Dang Wai Min Jhu Yun Dong De Ji Shi," reprinted in Chen Fang-ming eds., *Tai Wan Yi Shi Lun Jhang Siuan Ji*. LA: Taiwan Publisher.

Cheng Nan-ron, ed. (1987). *You Ren Zhu Zhong Tai Wan Zai Jue*. Taipei: Cheng Nan-ron.

Cheng, Fu-shin, ed. (1987). *Mei Li Tao Hou De Tang Wai*. Taipei: Chen Fu-Shing.

Cheng, Tung-jen (1989). "Democratizing the Quasi-Leninist Regime," in *Taiwan, World Politics*, XII-4.

Cline, Ray S. (1989). *Jiang Ching-Kuo Remembered: The Man and his Political Legacy*. Washington, D.C: United States Global Strategy Council.

Chou Yu-kou (1994). *Li Deng-hui 1993*. Taipei: Wu Shi.

Cohen, Bernard (1987). "The Census, Social Structure and Objectification in South Asia," in *Anthropologist among the Historians and other Essays*. Delhi: Oxford University Press, pp. 224–253.

Crag Calhoun (1990). "Introduction," in *Habermas and the Public Sphere*. Cambridge, Massachusetts and London, England: The MIT Press.

Cromaroff, Jean (1985). *Body of Power Spirit of Resistance: The Culture and History of a South African People*. Chicago and London: The University of Chicago Press, pp. 680–681.

Deng Wei-siang (1981). "Min Jhu-Dang-wei Sin Sheng Dai De Shi Ye," in *Politicians*, vol. 5, April 16.

Fasseur, C. (1994). "Cornerstone or Stumbling Block: Racial Classification and the Late Colonial State in Indonesia," in *the Late Colonial State in Indonesia: Political and Economic Foundations of the Netherlands Indies, 1880–1942*. Leiden: KTTLV Press, pp. 31–56.

Feher, Frenc, ed. (1990). *The French Revolution and the Birth of Modernity*. Berkeley: University of Berkeley Press.

Foucault, Michel (1976). *The Archaeology of Knowledge*. New York: Pantheon Books.

——(1977). *Discipline and Punishment: The Birth of Prison*. New York: Pantheon Books.

Gold Thomas B. (1986). *State and Society in the Taiwan Miracle*. New York: M. E. Sharp, Inc.

Gramsci, Antonio (1971). *Selection from the Prision Notebooks*. New York: International Publishers.

Habermas, Jurgen (1991). *The Structural Transformation of the Public Sphere: An Inquiry into a Category of Bourgeois Society*. Cambridge, Massachusetts: The MIT Press.

——(1992) "Further Reflections on the Public Sphere" and "Concluding Remarks," in Calhoun, Crag, ed., *Habermas and the Public Sphere*. Cambridge, Massachusetts and London, England: The MIT Press.

Hobsbawm, Eric and Terence Ranger, eds. (1983). *The Invention of Tradition*. Cambridge: Cambridge University Press.

Hou Li-Chou (1977). "Siang-tu wen-siue Lun Jheng De Ci Yin Yu Jiao Dian," in *Siang tu Wu Ai*. Taipei: Bo-Siue Publisher.

HHsiao, Michael ed. (1989). *Tai Wan De Sin She Hui Yun Dong*, Taipei: Chu-Lieu.

Hsu gui-feng (1990). *Tai WanJheng Dang Ri Ji 1989*. Taipei: Chi-hai.

Huang Jhao-tung (1994). "Jhan Hou Tai Wan Du Li Yun Dong Yu Min Zu Zhu Yi De Fa Zhang," in Shih Cheng-feng, ed., *Tai Wan Min Zu Zhu Yi*. Taipei: Cian-wei Publisher, pp. 195–236.

Huang, Jia-shu (1994). *Guo Min Tang Zhai Tai Wan*. Taipei: Dai-Ching.

Huang De-Fu (1992). *Min Jhu Jin Bu Dang Yu Tai Wan Di Qu Jheng Jhi Min Jhu Hua*. Taipei: Shi-Ying Publisher.

Joseb, Bob (1990). *State Theory: Putting Capitalist States in Their Place*. University Park, Pennsylvania: The Pennsylvania State University Press.

Kuo Tien-shen (1984). "Sin Wei Ji Yu Sin Si Wang: Siang-tu wen-siue Lun Jhan Hou Tai Wan Wen Tan Fa Jhan De Kao Cha," in *Independence Evening Post*, July 30.

Kuo Yi-Ge (1984). "Tai Wan Li Shi Yi Shi Wen Ti," *Taiwan Times*, March.

Guang Hua Editors (1979). *The Republic of China is on the Move: President Jiang Ching-Kuo's First Year in Office Shaped the Policies that will Decide the Nation's Destiny*. Taipei: Kwang Hwa Publishing Co.

Lai Shi-yu ed. (1987). *Shen Pan Kuo Ming Tung*. Taipei: Taiwan Literature.

Li, Siao-Feng (1987). *Tai Wan Mian Jhu Yun Dong 40 Nian*. Taipei: Independence Evening Post Publisher.

Li Deng-hui (1994). "Sheng Wei Tai Wan Ren De Bei Ai," in *Independence Evening Post*, April 30.

——*De I Ciang Tian Li Deng-hui's 1000 Days: 1988–1992*. Taipei: Mei-Ten Publisher.

Li Siao-fong (1987). *Tai Wan Min Jhu Yun Dong Si Shi Nian*. Taipei: *Independence Evening Post* (press).

Liang Jin-feng (1984). "Wo De Jhong Guo Shi Tai Wan," in *The Progressive*, vol. 13, June.

Li Ao (1987). *Jiang Chin-Kuo Shi Tai Wan Ren*. Taipei: Jin-long.

Li, Dai (1987). *Taiwan Tang Wai Yung Dung*. Hong Kung: Kuang Chiu Ching.

Li Wong-tai (1984). *Dang-Wuai Zai Chih Fa Jhang Shi Lue*. Taipei: Ba Shi Nien Dai.

Li Yong-chi (1991). "Li Nian De Jian Chi Yu Shi Jian De Kuan Rong," in *Lin Jhuo-shui Wen Ji*. Taipei: Cian Wei.

Lin Jheng-jie and Chang Fu-chung (1978). *Ciuan Jhu Wan Shui*. Taipei: Kui-Kuan.

Lin Jheng-jie (1981). "Tai Wan Min Jhu Yun Dong Jiu Wen-Dang Wai Sin Sheng Dai De Dao Lu," in *Ling Wei Tangwei*. Taipei: Lin Jheng-jie, pp. 104–133.

Lin Jien (1987) "Shi Gai Ge Ti Jhi, Hai Shi Jian Ji Gu Tou? Lin Jheng-jie Tan *Mei Li Dao Hou De Tang Wai*," in Cheng, Fu-shin, ed., *Mei Li Dao Hou De Tang Wai*. Taipei: Chen Fu-Shing, pp. 81–104.

Lin Shi-yu (1981). "Tai Wan Jheng Zhi Bian Qian De Gui Ji," in *The Cultivists*, vol. 1, p. 35.

——(1982). "Lun Dang Cian Liang Da Gong Shi Wen Ti," in *The Cultivist*, vol. 21.

Lin Jhuo-shui (1983). "Fang Ci Li Nian Bi Ran Shang Shi Shi Li," in *The Cultivists*, vol. 17.

——(1984). *Wa Jie De Di Guo*. Taipei: Bo-Kuang.

——(1989). "He Tong Jhi She Jhang Kai Wen Hua Jung Ti Jhan," in *New Time Tribunes*, vol. 6, reprinted in 1991, *Jheng Za De She Hui Yu Wen Hua*. Taipei: Cian Wei, pp. 123–206.

——(1992). "Tai Wan Jian Guo De Ren Wen Si Kao_Lin Jhuo-shui Wen Ji Yan Tao Hui," in *Shi Jie Yu Guo Jia Wen Hua Zung Zu*. Taipei: Cian Wei, pp. 167–182.

Lu, Shiu-lian (1991). *Chong Shen Mai-li Tao*. Taipei: Independence Evening Post Publisher.

Marshall, T. J. (1965). *Class, Citizenship, and Social Development*. Garden City, N.Y.: Anchor.

Mehta, Uday (1990). "Liberal Strategies of Exclusion," *Politics and Society*, vol. 18, no. 4, pp. 427–454.

Pender, Chr. L. M., ed. (1977). "The Impacts of Liberalism on Dutch Colonial Policy, 1848–1900," in *Indonesia Selected Documents on Colonialism and Nationalism, 1930–1942*, pp. 31–43.

Peng Rui-Chin (1991). *Tai Wan Shin Wen Sihue Yun Dong 40 Nian*. Taipei: Independence Evening Post Publisher.

Peng Ping-kuang (1977). *Dang Cian Wen-Siue Wen Ti Zung Pi Pan*. Taipei: Ching-Xi Literature Association.

Rafael, Vincente (1993). "White Love: Surveillance and Nationalist Resistance in the U.S. Colonization of the Philippines," in Amy Kaplan and Donald Pease, eds., *Culture of United States Imperialism*. Durham: Duke University Press, pp. 185–218.

Said, Edward W. (1978). *Orientalism: Western Representations of the Orient*. London: Routledge & Kegan Paul.

Sie Chang-ting (1983). *Dang-wei-dang*. Taipei: Voice of Awareness.

Sie Li-fa (1987). "Chong Er Er Ba Shi Jian Kan Tai Wan Jhi Shi Fen Zi De Li Shi Mang Dian," in *Min Jin Bao Zhou Ka*n, March 11.

Shu Kwei-feng (1990). *Tai Wan Jheng Dang Ri Ji 1989*. Taipei: Shi-Hei.

Skocpol, Theda and Margaret Somers (1980). "The Uses of Comparative History in Macrosocial Inquiry," *Comparative Study of Society and History*, vol. 22 (2), 1980, pp. 174–197.

Stoler and Cooper (1996). "Between Metropoles and Colony: Rethinking an Agenda," in *Tensions of Empire: Colonial Cultures in a Bourgeois World*. Berkeley: University of California: Berkeley Press.

Stoler, Ann (1992). "Sexual Affronts and Racial Frontiers: The Cultural Politics of Exclusion in Colonial Southeast Asia," in *Comparative Studies in Society and History*, vol. 34, no. 3, pp. 514–551.

Tilly, Charles (1984). *Big Structure Large Process Huge Comparisons*. New York: Russel Sage Foundation.

Tsai Yi-min (1984) "Shi Lun Chen Ying-Zen De Jhong Guo Jie," *The Progressive*, vol. 13, June.

Turner, Bryan S., ed. (1993). *Citizenship and Social Theory*. London: Sage Publication.

Vincente, Rafael (1993). *Contracting Colonialism: Translation and Christian Conversion in Tagalog Society under Early Spanish Rule*. Durham: Duke University Press.

Wachman, Alan M. (1994). *National Identities and Democratization*. Armonk, New York and London, England: M. E. Sharpe.

Wang Wun-sing (1978). "Wang Wun-sing Jiao Shou Tan Siang-tu wen-siue De Kung Yu Guo" and "Wang Wen-hsing Jiao Shou De Jing Ji Guan Yu Wen Hua guan," in Peng, Ron-li and Hsiao, Kuo-ho, eds., *Zhe Yang De Jiao Shou Wang Wun-sing*. Gaoxiong: Dun-Li Publisher, pp. 15–49.

Williams, Raymond (1977). *Marxism and Literature*. New York and Oxford: Oxford University Press.

Wu Jhong-liu (1984). *Wu Hua Guo*. Taiwan.

Wu Rey-ren (1994). "Jhui Ciou Ming Yun Gong Tong Ti-Ren Min Zi Jiou Ciuan Yan Yu Jhang Hou Tai Wan Gong Min Min Zu Jhu Yi," in *Tai Wan Ren Zi Jiou Suan Yan San Shi Jhou Nian Ji Nian Yan Tao Hui*. Taipei, September 16.

Yie Shi-tao (1979). "Tai Wan Siang-tu wen-siue Shi Dao Lun," in *Tai Wan Siang tu Zuo Jia Lun Ji*. Taipei: Yuan-Jing Publisher, pp. 1–25.

Yie Ya-Ming (1983). "Yi Shi Yu Chun Jhai-Jhai Lun Tai Wan Yi Shi," in *The Cultivists*, vol. 5, August.

Ruo Lin Jhen Chang (1992). *Tai Wan: Fen Lie Guo Jia Yu Min Jhu Hua*. Taipei: Yue Dan Publisher.

Yu Tian-chong (1977). "Wen-siue Wei Ren Sheng Fu Wu," in *Sia-Ch'ao*, vol. 3, no. 2.

Chapter 4

Hand Puppet Theater Performance: Emergent Structures and the Resurgence of Taiwanese Identity

Sue-mei Wu

Introduction

Hand Puppet Theater (*Putaihsi*) is a traditional performance art in Taiwan in which small puppet figures are manipulated by hand.[1] The great skill of the puppeteers and the sophistication of the puppets, some of which are capable of facial expressions, allow hand puppet theater to be very entertaining, effectively delivering both action scenes and dramatic scenes.

Originally developed during the Ming Dynasty (1368–1644), hand puppet theater was introduced to Taiwan about two hundred years ago by immigrants from Fukien province, where the hand puppet theater tradition had developed to a more sophisticated level than elsewhere in China (Liu 1990: 17–20; Ch'en 1991: 183). During its development in Taiwan, hand puppet theater has become intertwined with Taiwan local customs and practices, and later became a valued aspect of Taiwanese culture.

According to Liu (1990), there have been several stages in the development of hand puppet theater in Taiwan. When hand puppet theater first came to Taiwan during the Ch'ing dynasty (1644–1911) performances were mainly accompanied by music of the 'south-pipe' (*nankuan*) variety. This type of music was produced by stringed instruments and reflected the soft and elegant singing and movements of the hand puppet theater of that time. The stories were emotional love stories, whose slow pace was accompanied by elegant literary dialogue and soft singing. Over time, however, story lines based on popular serial novels and folk stories gained in popularity. These tales, based in the traditional martial world, were characterized by a fast pace and frequent combat scenes. They were accompanied by a different style of music, called the "north-pipe" (*peikuan*) variety. Instead of the string instruments of the *nankuan* style, the *peikuan* style utilized mainly percussion instruments, were better suited for the fast pace and frequent combat scenes of the new story lines. This style of puppet theater became more popular in Taiwan (Liu 1990: 56–61).

Hand puppet theater developed into a popular means of folk entertainment in Taiwan and experienced several cycles of relative popularity and decline. During the mid-1970s and 1980s, following a period of great popularity, hand puppet theater was in decline as a result of competition from more modern, technologically sophisticated forms of entertainment, as well as political policies that discouraged the use of Taiwanese language and displays of folk religion. Hand puppet theater in Taiwan was in danger of becoming what Dell Hymes terms "performance in a perfunctory key," that is, a performance undertaken out of a sense of cultural duty or traditional obligation, but offering little pleasure or enhancement of experience (personal communication with Hymes, cited in Bauman 1977: 26–27).

This chapter focuses on performance forms of the 1990s, a time period in which hand puppet theater enjoyed renewed popularity. It has adapted to the competition from modern forms of entertainment by becoming flashier and technologically sophisticated. In addition, there has been increased interest in preserving and promoting Taiwan hand puppet theater in its "pure" traditional form. Overlying both these trends is another important factor contributing to the resurgence of interest in hand puppet theater in Taiwan; hand puppet theater has become intertwined with the "Taiwanese" movement, which advocates an increased awareness and promotion of Taiwanese language, popular culture, and folk religion. Hand puppet theater is considered one of the most representative traditional Taiwanese performance arts and it seems to be evolving into an important symbol of Taiwanese culture, a symbol that has surpassed the realm of the arts and appeared in broader political discourse, as evidenced by its use in recent election campaigns in Taiwan.

This chapter also explores these recent trends in the development of Taiwan hand puppet theater. After a brief description of traditional Taiwan hand puppet theater and an overview of the political, economic, and cultural setting from which modern forms of hand puppet theater arose, I will introduce four contemporary types of hand puppet theater performance that are common today, including open-air performances at temple festivals, hand puppet theater adapted for television, exhibitions of traditional-style hand puppet theater for the purposes of preservation and entertainment, and a new hybrid form of hand puppet theater used as a tool for cultural education. Finally, a sample of fieldwork from an actual open-air performance will be presented to illustrate some of the changes that are occurring.

Traditional Hand Puppet Theater Performance

Traditionally, hand puppet theater troupes were sponsored and organized by folk organizations. Troupes were invited to perform at temple festivals held in honor of local deities and at auspicious occasions such as weddings, births, and promotions. The main purpose of hand puppet theater performance was to thank and entertain local deities. However, due to the lack of entertainment options in Taiwan's traditional agricultural economy, hand puppet theater also served as a popular means of folk entertainment.

A traditional hand puppet theater troupe is usually composed of two groups, the puppeteers and the musicians. The puppeteers typically number at least two, the main puppeteer and an assistant puppeteer. The director of the troupe usually serves as the main puppeteer and is in charge of manipulating the main puppets, performing the difficult scenes, singing, and narrating. The assistant puppeteer manipulates other puppets in coordination with those of the main puppeteer, changes the costumes of puppets, and takes care of the stage setup. The relationship between the two puppeteers is typically that of master and student. Frequently, the master trains his sons to succeed him as puppet master. As for the musicians, traditionally there are six to seven who sit backstage and play the instruments, including the Chinese two-stringed violin, gongs, cymbals, drums, castanets, and wind instruments. The drummer is in charge of directing the musicians during the performance.

The stage is a miniature version of a traditional temple, elaborately carved from fine wood. The stories to be performed are generally selected from popular folktales and historical novels, with the martial world being especially well-liked.

The language of the traditional performance has a very literary diction, typically including some poetry and literary idioms. Taiwanese is the main language of the performances.

Economic and Political Changes Affecting Taiwan Hand Puppet Theater

As technology spreads and the world moves toward one unified consumer culture, entertainment is becoming ever more technologically advanced and sophisticated. Thus, traditional forms of entertainment must compete with more modern distractions, such as television, movies, video games, and the like. In Taiwan, this transformation has been quite dramatic as industrialization and economic growth have occurred at a strikingly fast pace. Similarly, social and political changes in Taiwan have also been rapid, and contemporary forms of Taiwan hand puppet theater show their influence.

Taiwan has emerged as one of the world's fastest-growing economies, and its people now have access to many new forms of entertainment such as television, laserdisc movies, karaoke, and video games. The wide variety of entertainment options has created competition, that has presumably contributed to the reduction in demand for open-air hand puppet theater performances. This reduction in demand coupled with the rising wages and prestige of white collar employment in the commercial sector has caused a decrease in the number of professional hand puppet theater performers; current performers are tempted to seek more economically rewarding employment, while students do not want to give up their opportunities in the commercial sector to become puppeteers. As the number of professional performers has declined, performances have had to become more efficient. Whereas a traditional troupe consisted of around six musicians and two puppeteers who sang and performed dialogue live, it is now common to see the use of taped music and dialogue so that the whole performance can be handled by one puppeteer and one assistant. Moreover, many professional puppeteers now find it necessary to have second careers in order to support their families. For example, the puppeteer whose performance is described in the fieldwork section, works part-time selling chickens in a traditional market in an alley near his home.

Economic difficulties faced by hand puppet theater performers are not solely the result of economic forces. During the past several decades, Taiwan's rapidly changing political environment has also left its mark. After World War II, Taiwan, that had been a Japanese colony since 1895, was turned over to the Chinese government which was controlled by the Nationalist party (Kuo Min Tang). After being defeated by the communists on the mainland, the KMT retreated to Taiwan, along with what was left of their army. In total, between one and two million mainlanders fled to Taiwan, adding to Taiwan's population at that time of about six million (Wachman 1994: 21). The KMT hoped to use Taiwan as the base for the eventual reconquest of the mainland. This goal had a strong effect on their policies. From May 19, 1949, until July 15, 1987, Taiwan's constitution was suspended and it was ruled under what was effectively martial law. The KMT allowed no dissent, and the formation of competing political parties was forbidden. During this period of one-party rule and martial law, political emphasis was placed on eventual reunion with the mainland.[2] To support this goal, the KMT made an effort to socialize the population of Taiwan. Perhaps suspicious of the loyalties of the Taiwanese, who had lived under Japanese rule for fifty years, the KMT attempted to "re-sinicize" them. In schools, Mandarin was the only language permitted. Lessons in history and geography focused on the mainland, and the gentry culture of the mainland elite was presented as the true national culture (Wachman 1994: 40). Expressions of local Taiwanese culture were discouraged. In addition, The KMT had traditionally been opposed to popular folk religion, even on the mainland, seeing it as a waste of resources that could be put to better use in building the nation. For these reasons, the KMT attempted to suppress Taiwan folk religious festivities, which they considered to be superstitious, ostentatious, and wasteful, through a number of campaigns and policies (Bosco 1994: 396). One such policy, the "frugality policy," was announced in 1952. The government prohibited large temple festivals and encouraged limitation of expenditures associated with smaller folk religious activities (Liu 1991: 72). One result of this policy was a great reduction in the open-air hand puppet theater performances that had been associated with such popular folk religious festivals.

Recently the political situation in Taiwan has become much more open. Martial law was lifted by Chiang Ching-kuo in 1987, and there is now a multiparty system and greater freedom of expression.[3] The government has also relaxed its repression of Taiwanese culture. The *Free China Review*, which is published by the Government Information Office, now reports on Taiwan's cultural and economic trends, whereas it had focused on the mainland before the 1980s. In 1990, the magazine focused solely on Taiwan affairs, stressing the value of Taiwan's traditional arts as part of the Chinese historical tradition. In addition, the Art Heritage Awards, which were established in 1985 to increase the prestige of traditional folk arts, have been broadcast live on television since 1990 (Bosco 1994: 400). This is in marked contrast to the government restrictions on Taiwanese language programming that had already been in effect, and which will be discussed in more detail in a later section.

The Resurgence of Taiwanese Identity

It is out of this background of political and cultural repression, followed by a period of cultural, political and societal liberalization, that a strong sense of Taiwanese identity has emerged. Some commentators deny the existence of a distinct Taiwanese culture or identity because Taiwanese religion, language, and other cultural practices are similar to those of Fukien Province, from where had come the majority of the first wave of immigrants to Taiwan. However, two recent commentators, Alan Wachman and Joseph Bosco, argue that identity need not be based on observable cultural differences, but rather is a sense of group consciousness that is invented or created in response to social and political forces. Wachman and Bosco see Taiwanese identity as having grown out of the experience of being separated from the rest of China and being ruled by foreigners such as the Dutch and the Japanese. Later, the KMT's political dominance and repression of Taiwanese language and folk culture reinforced the Taiwanese feeling of being distinct, which was actually counter to the KMT's goal of bringing the Taiwanese "back into the fold" as it were (Wachman 1994: 40–44; Bosco 1994: 392–397). Indeed, elements of culture that were repressed by the KMT began to serve as symbols or markers of Taiwaneseness. Bosco points out that after the KMT restrictions on popular religious festivals, these festivals and the display of the pig that was their common feature became symbols of Taiwaneseness (Bosco 1994: 396). Since hand puppet theater was also closely tied to these festivals, this was probably also the beginning of hand puppet theater's role as a symbol of Taiwaneseness.

So far we have discussed one form of Taiwanese identity, namely the sense that earlier immigrants to Taiwan have of being distinct from those who came to Taiwan at the end of World War II. Bosco feels that contact with the mainland in the last few decades has created a new form of Taiwanese identity, a sense shared by all residents of Taiwan, regardless of the timing of their arrival or home province on the mainland, of being distinct from residents of the PRC. Bosco argues persuasively that this second influenced by the first sense of Taiwanese identity, and is becoming stronger as contact with the mainland increases, highlighting differences between the mainland and Taiwan (Bosco 1994: 393–394). This sense of identity has also grown out of political necessity, as a more open, democratic political process motivates politicians to search for ways to connect with citizens, the majority of whom identify themselves as Taiwanese. Indeed, Bosco points out that it was the KMT's local politicians themselves who undermined the restrictions on popular folk religion, because popular religious ceremonies and organizations are important forums for building local political support (Bosco 1994: 396–397). Such considerations may also explain the embracing by the KMT of elements of Taiwanese folk culture that they once despised and attempted to suppress.

More evidence of this trend can be seen from the use of the Taiwan hand puppet theater as a symbol to bolster the Taiwanese identity of candidates in recent election campaigns. For example, the first ever direct election of the governor of Taiwan Province in history occurred on December 3, 1994. The winner, Sung Ch'u-yu who is mainland-born, made use of hand puppet theater puppets in his campaign. Moreover, in the first direct presidential election, March 23, 1996, President Lee Teng-hui expressed his deep concern for the Taiwanese people by supporting and promoting most of the Taiwanese folk arts, of which hand puppet theater was most closely involved in his election campaign. For example, President Lee's name and seal appeared in the foreword to a set of documentary videotapes about the puppet theater of Li T'ien-lu, one of the great masters designated as a "National Treasure." These tapes are part of the series called "Let the Tradition Continue" produced under the guidelines and support of the Government Information Office of the Executive Yuan.

Recently, the newly elected Taipei Mayor, Ma Ying-chiu, portrayed himself as the incarnation of a well-known hand puppet theater puppet character, Shih Yen-wen, during his campaign (cf. *China Daily News* July 29, 1998). Shih Yen-wen is one of the most well-known and beloved of puppet characters, known by young and old as a knowledgeable scholar with martial skills who constantly fights for what is right and strives to create a better life for common people.

As a result of the growing power of those two forms of Taiwanese identity, Taiwanese language and culture are experiencing a renaissance. The study of Taiwanese language and folk arts is now supported by the KMT. In September 1998, the study of Taiwanese language became part of the official school curriculum beginning with the third grade of elementary school in Taipei city (cf. *World Daily News* July 2, 1998). Hand puppet theater, now promoted as one of the great achievements of Taiwanese culture has begun to function as its representative symbol. Recently, I stumbled upon anecdotal evidence of this new role of hand puppet theater when I noticed that the cover of a tape of popular Taiwanese songs I purchased was decorated with pictures of various hand puppets. Thus, the puppets are also identified with other strong expressions of Taiwanese culture, namely Taiwanese lyrics and music. In another example, at the 1996 Asian Festival at Ohio State University, the Taiwan Students Association opted to stage a hand puppet theater performance to represent their culture. It is significant that these relatively young Taiwanese selected hand puppet theater from among the many things that they could do to introduce Taiwanese culture to other students. We will see more evidence of this role in the following analysis of the various forms taken by contemporary hand puppet theater.

Emergent Structures in Contemporary Hand Puppet Theater Performance

In this section we will examine emergent structures in hand puppet theater in order to see how the form of the performance is being adapted to the rapid economic, social, and political change in Taiwan. It is important to recognize that the emergent structure of a performance, in Bauman's conception of the term, does not simply refer to what is new, or what differs from tradition, but to how each different feature of the performance is situated on the continuum that runs between the poles of the completely novel and completely fixed. This structure, Bauman states, "is of special interest under conditions of change, as participants adapt established patterns of performance to new circumstances" (Bauman 1977: 42). Under the conditions of change in Taiwan, we see two different trends in the emergent structure of contemporary hand puppet theater performance. One is the adoption of new technology and conventions, both within the traditional loci of performance and also in new contexts of performance. The second is the renewed emphasis on preserving traditional forms, but within different performance contexts and to serve different social purposes. These two trends are evident in the different forms of contemporary hand puppet theater described in the following section.

Open-air Performances During a Temple Festival

Many troupes still perform at the traditional locus of hand puppet theater performance, that is, during a temple festival at a temple, which also serves as a community center of activity. However, open-air performances reflect the pressures on hand puppet theater as it attempts to compete with more modern forms of entertainment, as well as the legacy of KMT campaigns against displays of popular folk religion. The stages used are often not as elaborate as traditional stages, which can be very impressive works of craftsmanship. Although painted elaborately, the new stages are fairly simple in construction, allowing them to be portable. This reflects the declining popularity of open-air hand puppet theater and the resulting economic pressures. There are fewer performers, and as performances are not as lucrative as they once were, performers are required to travel in order to put on as many shows as possible. Nowadays, many utilize a truck bed as the base for a highly portable stage. Further, reflecting the economic pressures on modern hand puppet theater performers, we see that whereas a traditional hand puppet theater performance requires several performers, including a puppeteer, assistant puppeteer, and several musicians, the performance described in the fieldwork section utilized only two people, the puppeteer and his wife. In order to put on the show with so few performers, taped music and dialogue are used enabling the performers to concentrate on manipulating the puppets. To make the scenes more eye-catching, they utilize multicolored flashing lights, and smoke produced from aerosol cans. In addition, because the temple festivals do not last very long, the plots for the performances cannot be too complicated.

The audiences for open-air performances appear to consist mainly of older folks and young parents with their children. Presumably, the older folks have enjoyed hand puppet theater performances since their youth, when it may have been the best existing form of entertainment, and the young parents may be introducing their children to what they consider an important facet of Taiwanese culture.

Hand Puppet Theater Performance Adapted for Television

The medium of television is one of the new contexts of hand puppet theater performance, and certainly the most popular today. It is an adaptation that has arisen out of both economic and political trends. As Taiwan's economic situation improved, more and more families owned television sets, giving television broadcasts an even larger potential audience. As the political situation became more open, Taiwanese language programming and an increase in the number of TV stations was permitted, allowing televised hand puppet theater more than just token exposure on television.

When television was first aired in Taiwan in 1962, Taiwanese programming was very popular (Wachman 1994: 53). The first successful televised hand puppet theater appeared on the scene in 1970, with a very popular series created by Huang Chun-hsiung who is the son of the famous puppet master (and designated "National Treasure") Huang Hai-tai (*Sinorama* 1998: 147). This series introduced several reforms: Puppets were made larger, popular music was used instead of the traditional Chinese instruments, the stage was replaced by a painted background, and performances were accompanied by neon lights and electronic special effects. This new form of hand puppet theater was called *Chinkuanghsi* ("Gold Light Show") (Liu 1990: 74–77). In response to the popularity of Taiwanese programming, in 1972 the government limited Taiwanese programs to half an hour at lunch and half an hour at night (Wachman 1994: 53). Four years after it had begun, Huang's program was taken off the air because it "disrupted the normal routine of agricultural and industrial work and thus shook the foundations of the nation" (*Sinorama* 1998: 147). In 1976 the government went one step further, requiring that all programs had to be broadcast in Mandarin (Wachman 1994: 53). There were some experiments using Mandarin for televised hand puppet theater, but they were unsuccessful (Liu 1990: 14; *Sinorama* 1998: 147).

In recent years, however, the number of TV stations in Taiwan has increased from the three commercial stations closely controlled by the KMT—Taiwan Television Enterprise (TTV), China Television Company (CTV), and China Television System (CTS) to over sixty stations, including cable and satellite television. Illegal cable stations that included Taiwanese programming had begun broadcasting in 1976, and then, following their enormous popularity, the government passed a cable television law in 1993 which detailed a legal framework for authorized cable systems. The three commercial stations affiliated with the KMT were also joined in by a fourth commercial station, Formosa Television Corporation which is affiliated with the opposition Democratic Progressive Party (DPP). Formosa TV focuses on Taiwanese language programming (*The Republic of China Yearbook 1997*: 284–286).

In 1995, Huang Chun-hsiung's sons, Huang Ch'iang-hua and Huang Wen-tse, established the *Pili* ("High Energy") satellite station to broadcast their *Pili* puppet theater productions, as well as other popular Taiwanese programs. The *Pili* puppet theater is a descendent of Huang Chun-hsiung's early televised puppet theater, and is now very different from traditional hand puppet theater. On television, hand puppet theater is separated from its culturally stated purpose of thanking and entertaining the deities, and exists predominantly as an entertainment form. Hand puppet theater programs on television are not confined to a stage, but can have elaborate sets. The medium also allows a high degree of technological advancement and audiovisual special effects. This facilitates the addition of supernatural elements into the plots. In addition, on television, hand puppet theater is not performed live, so it may be edited time and time again in search of perfection. This allows performances to include more spectacular and complex moves, especially in the combat scenes. For the music, traditional instruments have been replaced by recorded popular Taiwanese music. Finally, separating hand puppet theater from traditional contexts of performance also adds freedom and regularity to the scheduling of the shows. A nightly viewing audience facilitates complex plot structures that can unfold episode after episode. This contrasts to the situation at temple festivals or other traditional contexts of performance, where the entire plot line must be completed in one or just a few shows.

On television, hand puppet theater, especially the *Pili* series, has subsequently become a highly successful form of commercial entertainment. Some characters from the series have become very popular and are loved by people from many different ages and social classes. New hybrids have arisen out of this trend. For example, with the popularity of video games and Japanese comic-book series in Taiwan, the famous puppet characters and stories of the *Pili* TV series have been adapted as video games and comic book series. The Huang brothers have even built a movie studio and produced a feature film. The hand puppets, like human stars, attend press conferences to drum up fan interest. The *Pili* hand puppet theater series has targeted the 18–35-year age group who are comic book, adventure novel, and/or video game fans. Televised hand puppet theater has become so popular and successful that college students can be heard avidly discussing the characters and plots on campus. Now there are also web sites for information and for chatting about hand puppet theater characters and stories, and they have been receiving heavy traffic from the internet.[4] Further, they have opened a chain of *Pili* specialty shops in Taiwan, where fans can join the *Pili* fan club, and buy puppets, videos, posters, trading cards, notebooks, puzzles, and many other items.

This form of hand puppet theater is truly becoming uniquely Taiwanese, in Bosco's interpretation of Taiwanese as distinct from mainland Chinese. It was developed wholly in Taiwan, and Huang Wen-tse describes the language used as "a kind of Mandarinized Taiwanese because "some terms from Mandarin are not translated into Taiwanese according to their meaning, but simply have their written characters from Mandarin pronounced in the Taiwanese dialect. Other times, Taiwanese terms come out in Mandarin" (*Sinorama* 1998: 149).

Exhibitions and Education

As mentioned in the previous section, hand puppet theater is increasingly being seen as representative of Taiwanese local culture, both traditional and modern. This also leads to new contexts of performance, as various groups choose to stage traditional performances to showcase hand puppet theater as an important aspect of Taiwanese culture. Such performances, although aimed at preserving the traditional style of hand puppet theater, are new contexts of hand puppet theater performance, removed from the temple festivals, weddings, and other auspicious occasions where hand puppet theater performances traditionally occurred. The sponsorship of such performances has also changed. Traditionally, troupes were sponsored by community folk organizations and temples, and these organizations still sponsor performances such as the one described in the fieldwork section below. However, intellectuals have also begun to support this art through groups such as the Se-Den Society (*Hsi T'ien She*), a foundation set up by a group of Taiwan University scholars for the purpose of showcasing and preserving traditional Taiwan hand puppet theater. The Se-Den society takes its name from an acronym composed of the first parts of the names of Lord Hsi-ch'in and Marshall T'ian-tu, two ancient folk heroes. According to legend, both of them were steadfast patrons of the traditional theater, including hand puppet theater. Accordingly, they have since been considered the patron saints of the folk arts. The Se-Den society has established a full-fledged traditional hand puppet theater troupe, and has also carried out other work related to preserving and promoting Taiwan hand puppet theater in its traditional form. The goals of the Se-Den Society, summarized from their brochure entitled "A Briefing on Se Den Society" are listed here.

1. Preparation for the establishment of a Taiwanese traditional drama hall.
2. Establishment of a full-fledged, well-organized hand puppet theater troupe.
3. Compilation of a series of articles on the hand puppet theater puppets.
4. Compilation of an edition of memoirs of hand puppet theater dramatists, master performers, and senior entertainers.
5. Preservation of articles, materials, and other documents on traditional Taiwanese drama and its troupes.
6. Founding of scholarships for research on hand puppet theater.
7. Introduction of hand puppet theater into the campuses of various schools at various levels, either as a curricular or extracurricular activity, thus familiarizing the younger generation with the essence of our treasured Taiwanese cultural heritage.

8. Advocacy and encouragement of the revitalized acceptance of hand puppet theater and the sponsoring of the art form's public performances as well as Taiwanese traditional drama by local religious organizations and establishments.
9. Field investigations of the development and popularity of hand puppet theater throughout the Island.

We can see that this group is dedicated to preserving traditional hand puppet theater as a representative example of traditional Taiwanese culture. They have commissioned an intricate traditional stage, and since 1987 have sponsored an annual public performance of traditional hand puppet theater, complete with a full troupe of musicians and performers. The Hsiao Hsi Yuan troupe and I Wan Jan troupe are two other troupes dedicated to performing the traditional style of hand puppet theater. Their puppet masters are Hsu Wang and Li T'ien-lu respectively. They are both considered among the most talented puppet masters. Their performances are in great demand not only in Taiwan, but also in cultural centers throughout the world.

Taiwanese people seem to have longed to express their identity and now that there is no repression they are proud to display an art form they claim as their own. The KMT government has also co-opted this cultural strain into their campaign to preserve the Chinese cultural heritage. This has lessened the tension between the high culture first promoted by the KMT and the Taiwanese popular culture that they had considered as low culture (Bosco 1994: 397–400). One example of how the KMT has embraced hand puppet theater and associated it with the high cultural tradition on the mainland was the selection of the first exhibit for the new Cultural Gallery at the National Concert Hall in Taipei. The Council for Cultural Planning and Development of the Executive Yuan opted to organize "An Exhibit on Traditional Chinese Glove Puppetry" as the gallery's first show "in the hope that this exhibit will attract the public to search for its roots and come to know of Chinese traditional folk culture." See program for the exhibit held February 18–March 12, 1992, at the Culture Gallery, National Concert Hall. Note that both the title of the exhibit and the statement of its goal place hand puppet theater in the Chinese traditional folk cultural tradition, rather than promoting it as something uniquely Taiwanese. Another example is the treatment of Taiwanese folk arts in the *Free China Review* that was mentioned previously.

A New Hybrid Form of Hand Puppet Theater Used for Education

There are now experiments in using another hybrid form of hand puppet theater as a means to promote and educate people about Taiwan local history, legends, and geography. For example, the *Erhshui Ming Shihchieh Changchung chut'uan* ("The Erhshui Ming Shihchieh Hand Puppet Theater Troupe") is devoted to the use of hand puppet theater as a means to educate people about Taiwan local culture, history, legends, and geography and also to propagate it. The playwright is ethnologist Chiang Wu-ch'ang who resigned from his teaching position at the National Art Institute in order to devote himself full-time to doing field work and writing the scripts for the shows. The director of the troupe is Mao Ming-fu. Students from Providence College volunteered to assist in performing the show. They claim that theirs is the first troupe that is truly concerned about educating and guiding Taiwan people to better understand the history and legends of their native land. The first experimental show, entitled *Erhpa Shui Feng Yun* ("Unpredictable changes of events at the Erhshui area"), was performed on March 6, 1997, in Changhua County in central Taiwan. It describes the toils and struggles of Changhua's ancestors to open up the barren lands for farming, including as well some touching historical events such as ethnic fighting and the events of February 28, 1947 (*Erh Erh Pa*). Since the show is not set in ancient China, the troupe created new puppets and costumes modeled on the contemporary clothing of their subjects (*Central Daily News*, March 7, 1997).

This is an ironic turnaround. Hand puppet theater was used by the Japanese and by the KMT in earlier propaganda campaigns. During the colonial period, the Japanese did not place major restrictions on expressions of Taiwanese folk culture. However, after the beginning of World War II the Japanese began placing restrictions on hand puppet theater and other forms of traditional performance. Puppeteers had a choice: they could either reform their performances or find something else to do. Many troupes, in order to survive, switched to Japanese-style puppets and themes that supported the cause of the Japanese empire (Liu 1990: 36, 68–69). After the Japanese left, the KMT used hand puppet theater to advance anti-communist and anti-Russian themes. The language used in such performances was Mandarin, and the puppets wore contemporary clothing (Liu 1990: 70–71). Now, the Taiwanese are using hand puppet theater in an attempt to reclaim their identity.

Field Study

This section presents a performance of the Chang Shuang-hsi, hand puppet troupe in Panchiao city in the suburbs of Taipei, Taiwan.

On March 21, 1996, an open-air hand puppet theater performance was scheduled in honor of the birthday of the local earth deity (*T'utikung*). The community wished to celebrate his birthday and thank him for the security and prosperity he had provided. The setting for the performance was the local temple. The temple is located in the community center, which also contains a daycare center, small playground, open-air platform stage, and some offices for dealing with temple and community affairs. An alley runs between the community center and the temple.

The Chang Shuang-hsi hand puppet theater troupe was invited to hold a performance in order to entertain and thank the local deity, as well as provide a community entertainment activity. On the day of the performance, the community center area was crowded with locals, spanning both the old and young generations. Incense and paper money were everywhere. Sacrifices of fruit and canned foods were displayed on the tables.

The puppeteer, Chang Shuang-his, arrived and began to set up the stage for two scheduled performances, one in the afternoon and one in the evening. After explaining her purpose, the interviewer received permission to conduct an interview and videotape the performance. The interview was conducted mostly in Taiwanese, with a little Mandarin, and the puppeteer exhibited a friendly and easygoing personality in answering the questions. Mr. Chang is in his early thirties and has been doing puppet theater for fifteen years, continuing the traditional family business to make his living. Though he has had some students in the past, he feels that since society is changing, young people would rather be white-collar workers than choose hand puppet theater as their profession. Although Chang's full-time students are very few, he has benefited from the Taiwan government's recent support for and promotion of traditional folk arts. Mr. Chang has been invited to teach hand puppet theater in elementary schools as an extracurricular activity. He has also performed for a television program.

Since open-air hand puppet theater is mainly performed during temple festivals, festival times are very busy for Mr. Chang. Today, for instance, he and his former students are all busy with performances. Like many modern troupes, Chang has replaced the musicians with recorded music and added some high technology special effects to his repertoire, including laser lights, CDs, and smoke spray. Tonight he will perform a well-known scene from the cases of the famous Judge Pao (*Paokung*).[5] Recently a series of Judge Pao episodes was very popular on Taiwan television and then on the Hong Kong video market. Regarding the story chosen, Mr. Chang commented that when performing for a TV series that can last for several months, a new story about the traditional Chinese martial world is usually created. When performing for the open-air stage, however, he prefers to select one episode from a well-known traditional serial story. Tonight's Judge Pao show is no exception. With a traditional popular story it is easy for the audience to recognize the characters. For example, the audience can recognize the puppet figure as Judge Pao from the half moon on his face. Also, with the time constraints of an open-air performance (performances last about an hour and a half), the audience can follow better a plot with which they are already familiar. In a long television series, many complicated plot twists can be introduced, much like in American soap operas.

Soon after the interview, Mr. Chang's wife, who served as a temporary assistant performer, arrived. After learning the purpose of the interview, she welcomed the interviewer and kindly displayed some puppet figures, which were stored in trunks. She held up one of the most elegant and complicated lady puppet figures for the camera. When asked if she had learned how to manipulate the puppets from her husband, she laughed and shyly explained that she had tried for several years but could not learn it. "It is too difficult," she said. Tonight, since Chang's regular assistant performers are busy with other shows, she will serve as a temporary assistant. She will guide some puppets through basic movements, change the puppets' clothing, set up the props, change the background lights, and spray smoke to create atmosphere for the combat scenes.

Audience and Surroundings

Around 7:30 in the evening at the community center, an audience of about fifteen people has gathered. Some are sitting on chairs provided by the community center, some on rocks under a tree, and still others are perched on their motorcycles which they have parked on the grounds. The audience consists mainly of elderly people smoking cigarettes and sitting near the stage, but there are also some children with their parents and a few young and middle aged adults. People are passing to and fro on foot and by scooter along the small alley which separates the community center and the temple grounds. A lot of locals are coming to worship at the temple festival, but few are assembling in front of the stage to watch.

Stage

The puppet stage is set up on the community's open-air stage, which is used for many different kinds of performance. The open-air stage is about twenty feet across, and the puppet stage, which looks like a small temple, is about fifteen feet wide and twelve feet tall. The puppet stage has two layers, the front face of the stage that has an opening through which the puppets can be seen, and a back face that serves as the background for the scenes. In general, the puppets enter the stage from the left and exit to the right. A screen a few feet behind the puppet stage has an oil painting of natural scenery, including some mountains, trees, and light blue water. The stage is lit by three overhead lights that can be switched to different colors. Before the performance begins, some props for the first scene have already been set up. There is a desk and two chairs. The audience can tell it is some sort of a formal room.

Illuminated by a light placed on the ground, the front of the on-stage structure bears the name and address of the performing troupe "The Wuchou Chang Shuang-hsi hand puppet movie and theater troupe" (*Wuchou Chang Shuang-hsi Putaihsi Tienying Chut'uan*), as well as the name of the director, Chang Shuang-hsi, in large characters. There are also two large painted dragons on either side of the structure facing each other wildly with their mouths open. The structure is painted to represent the eaves of a temple, giving the whole on-stage structure the appearance of a traditional temple. It is splendid and colorful in appearance.

Prelude

Before the show, background music is played, consisting of renditions of Taiwanese popular songs but without the lyrics. Suddenly, the background music is replaced with the sounds of thunder and heavy winds. The overhead lights begin to flash and an aerosol spray is used to make the stage appear smoky. The puppeteer uses the microphone to announce that today is the four-year birthday of the community's earth god (T'u-di Gung) and that his troupe is going to perform the "Judge Pao" in celebration. He closes by inviting the audience to watch the show.

The Show

The thunderstorm lasts a few seconds, then, the sound of drums and a voice crying out "Hoo, hoo, hoo" indicate that the show is on. As in

Peking and Taiwanese opera, the audience can usually recognize the characters from their movements, costume, and gestures. Sometimes the puppets perform gestures and movements to the music as a kind of introduction before they speak. For example, in the first scene tonight, the first puppet figure jumps out from the left entrance slot with some movements indicating his martial skills. He is Wang Ch'ao, one of the two famous aides-de-camp of Judge Pao. Wang Ch'ao stands steadily on the left of the stage. Judge Pao enters from the left entrance slot. With a dignified air, he walks around the stage a few times before seating himself at the table. Thus before any speaking has occurred, the audience can recognize at the very least that one of the figures is a fighter and the other some kind of official. After sitting on the chair, Judge Pao announces his title and official rank. With literary diction, he explains that he is on a mission from the emperor to collect tribute along the northern border. Then Judge Pao, followed by Wang Ch'ao, exits the stage through the right exit slot. For the second scene, the stage props are quickly replaced with two chairs to symbolize a room in a normal house. A bandit leader brandishing a huge knife enters the stage with some bandit-like gestures. The overhead bulbs flash and turn purple. Then the bandit speaks, identifying himself and his plan; he plans to kill Judge Pao when the latter passes through bandit territory.

Combat scenes are among the most common and fascinating in puppet theater. They involve intricate manipulation of the puppets, special effects, and a high degree of coordination among the puppeteers. As such, fighting scenes are among the most difficult to perform. In this scene, the bandit fights with Judge Pao's aide. On the stage, we see the two puppets circling each other before attacking, chasing each other back and forth, running, jumping, and flying around the stage. The weapons are flashing in the light, and the audience gains the impression of a fantastic battle. Once again, the overhead lights have been changing colors and flashing. The stage becomes smoky. The music becomes fast-paced and is accompanied by the sounds of clashing metal, further enhancing the fighting scene.

Observing the fighting scene from backstage, we see the main puppeteer controlling both puppets, one in each hand. His hands are moving about wildly and he is moving quickly around the backstage area. He must walk and turn quickly to keep pace with the frenzied music. The assistant puppeteer is busy controlling the overhead lights, making the smoke, and adding the laser lights. Suddenly, the music slows down. The puppeteer places the two puppets on the board before him then quickly grabs another puppet and makes it walk slowly across the stage. The fighting scene which lasted only one minute, has been changed to another scene. The quick, life-like movements of the puppets and the fast, coordinated movement of the puppeteers backstage are marvelous to observe.

Text Sample

The following is a sample translation of some of the text before and after one of the fighting scenes.

Tumbling River Rat (*Fanchiangshu* a nickname, hereafter TRR) enters the stage.

TRR: "It's said that 'the *han-hsiao* flower in bloom is fragrant all over the mountainside, and a person of jade in bloom can handle everything well,' hei, hei, hei . . . ugh . . . Can you recognize this old man here? If you don't recognize me, I'll report on myself again so you'll know. Tumbling River Rat—Chiang P'ing [he gives his nickname and then his name]. They say that 'one who can be a man of leisure but doesn't take advantage of it is like a sow picking up garbage.' That's just like me. Before, I couldn't get used to the life on Hsienkung Island, so I went to Kaifeng in Fengyang county to come to the aid of my brothers in arms. I stayed there in the capital for over two years. Now, when I'm going back to Hsienkung Island . . . (I run into this situation)."

[Tumbling River Rat encounters Judge Po's aide, Wang Ch'ao, fighting with the bandit and realizes that the famous Judge Pao must be nearby and in trouble. After Wang Ch'ao retreats because he has realized that the bandit is a superior fighter, TRR confronts the bandit. We rejoin the scene after the bandit has made fun of Tumbling River Rat's somewhat advanced age.]

TRR: "If you don't have any hair in your belly button . . . no hair there then I'll just pull out the hair that's hanging on your shoulder armor."

[They fight and TRR kills the bandit.]

TRR: "This old man Tumbling River Rat, Chiang P'ing . . . because of this one phrase of yours . . . you said that 'I'm such an old man yet still come out and die for everyone to see' . . . Your daddy here [read "I"] got really angry. I'm like the crocodile [a brand name] mosquito-killing incense . . . if I kill more it's still not killing enough. Now give me all of you to beat to death! They say 'If you want to catch bandits you need to catch the leader first.' Well I just killed the chief. This red-hair [the red-haired bandit chief] . . . beat to death and now his feet are sticking up in the air [common Taiwanese expression, similar to the English "dead as a doornail"]. Hngh! Your daddy might as well set fire to your mountain fortress and burn it down! And then I'll go and see Master Pao."

[TRR sees Judge Pao.]

TRR: "Ah, I pay my respects to Master Pao."

PAO: "So it's Tumbling River Rat. Many thanks for your hero's rescue."

TRR: "Ah, Master Pao has left the seat of government at Kaifeng in Fengyang county. Where is the Master going now?"

PAO: "I've been sent by his Majesty the Emperor on a mission to the north to collect tribute. This time it's fortunate that a hero like you was here to rescue me, otherwise the whole army would be in mortal danger. I hope that all heroes will join the ranks who are going to collect tribute. I wonder what you think about this?"

TRR: "Ah, Master Pao, you're asking me to join your ranks to go to the north and collect tribute? I accept! I accept!"

The text structure of Taiwan puppet theater is still very similar to the traditional standard. Taiwanese language is used and the plots usually concern the traditional martial world. The characters still introduce themselves to the audience in a manner that is familiar to us from many forms of traditional Chinese fiction and theater. For example, when a character takes the stage he typically states his name as well as something about his personality or character. This is a convention that is still used even by characters in the ultramodern *Pili* puppet theater television series.

Despite the persistence of the traditional style of text, there are still a few emergent structures that reflect hand puppet theater's competition with modern forms of entertainment, and self-consciousness of its role as a symbol of Taiwanese culture. One that we have already mentioned is that the text is often produced beforehand and delivered on tape. Due to this economic necessity, the performer certainly gives up some flexibility and ability to adapt the performance to the audience. This adds to the separation between the performer and the audience. Beyond this, puppeteers sometimes introduce bits of modernity into their text in the form of foreign words or phrases, that are inevitably anachronisms in a traditional Chinese context. These anachronisms presumably add to the entertainment value of the performance and create a bridge between the traditional art and modern life. For example, several brands of products used in Taiwan are mentioned in performances. In one scene, the knight hero, TRR, mentions that his attitude toward the bandits is just like "crocodile brand mosquito-killing incense." This product is a common one that everyone in Taiwan has used or heard about. I thought it was funny when I heard it mentioned in the performance. This device also localizes the performance text, as only natives of Taiwan or those who are very familiar with daily life in Taiwan would recognize the reference. I have noticed a similar device used in other performances as well. In one televised hand puppet theater performance, also set in traditional China, one character asks another how they will get somewhere. The reply he receives is "I guess we'll have to fly, since the Toyota isn't working."

Another interesting feature of the text structure of contemporary hand puppet theater performances is that they seem to be placing special emphasis on their "Taiwaneseness" by including many Taiwanese maxims, many of which are not in common use and can be understood only by the older generation. They do not have to be immediately understood, however, to serve their function of marking the performance as uniquely Taiwanese. The maxims are generally spoken by a wise old character. For example, in this text sample Tumbling River Rat uses two such maxims: "The han-hsiao flower in bloom is fragrant all over the mountainside, and a person of jade in bloom can handle everything well" and "One who can be a man of leisure but doesn't take advantage of it is like a sow picking up garbage." In fact, one can now find books for sale in Taiwan bookstores that list and explain these maxims. The author of one such book explained in the preface that his purpose for collecting and explaining the maxims was to preserve the richness of the Taiwanese language and cultural tradition, and to introduce readers to some of the wisdom to be gained from that tradition (Li 1993: 3–5).

Concluding Remarks

In the last two decades, hand puppet theater has had to struggle to compete in a world of ever-increasing entertainment options. However, this traditional performance art is both adapting to the competition, and being preserved and promoted in its traditional form. Contemporary hand puppet theater in Taiwan can be divided into four types: open-air performances, TV performances for entertainment, exhibition performances for preservation and education, and new experimental forms designed to educate people about Taiwan local history and geography. Of these four types, technologically enhanced performances are gaining in popularity, especially on television where one can now find hand puppet theater performances broadcast on any given day. The increased popularity of all forms of hand puppet theater has, of course, also been supported by the growing perception in Taiwan of hand puppet theater as a symbol of Taiwanese identity. Purists who wince at the technological gimmicks of the televised hand puppet theater performances need not fear that adaptation to competition will change hand puppet theater beyond recognition. Indeed, because of hand puppet theater's symbolic value, efforts to maintain it in its traditional form have been successful. Although the context of performance has changed from temple festivals and auspicious events to concert and exhibition halls, major efforts by groups such as the Se-Den Society and traditional troupes such as the Hsiao Hsi Yuan and I Wan Jan, have ensured that people in Taiwan and people around the world can still go to see a hand puppet theater performance complete with intricately carved wooden puppets wearing elegant traditional dress, highly skilled puppeteers who sing and produce the dialogue live for every performance, traditional musical ensemble, and traditional carved stage.

Notes

1. In general, there are three types of puppet theater in Taiwan, including *K'ueileihsi* ("string puppet theater or marionette theater"), *P'iyinghsi* ("shadow puppet theater") and *Putaihsi* ("glove or hand puppet theater"). *Putaihsi* (cloth-bag-play) is also called *Changchunghsi* (palm-inner-play) and *Hsiaolung* (small-basket).
2. Martial law had been in effect in Taiwan from May 19, 1949, until July 15, 1987, when the Emergency Decree was lifted by President Chiang Ching-kuo. The ban on the establishment of new political parties was lifted in 1989. (Cf. *The Republic of China Yearbook* 1997: 67–69, 674.)
3. Since the ban on new political parties was lifted in 1989, eighty-two political parties have been registered with the Ministry of the Interior (as of 1996). The three most significant parties are the Kuomintang (KMT), Democratic Progressive Party (DPP), and the New Party (NP). (Cf. *The Republic of China Yearbook 1997*: 99.) Taiwan has also shifted to democratic elections. Members of the National Assembly and Legislature have been chosen by direct popular election since 1991 and 1992, respectively, and the reelection of President Lee Teng-hui on March 23, 1996 was the first direct popular presidential election in Chinese history. (Cf. *The Republic of China Yearbook 1997*: 69.)
4. Under the Yam search directory, http://www.yam.com.tw/b5/cult/folk/tpuppet/, over thirty internet sites dedicated to hand puppet theater are listed. High Energy Puppet Productions also has their own site at http://www.highenergy.com.tw/. Links to College BBS stations for discussing hand puppet theater may be found at http://web.yam.com.tw/.
5. Judge Pao was an upright official known for his respect for the law. He was a famous character whose cases are presented in the Chinese traditional novel entitled *Chi'hsia Wui Paokung An* ("Seven Knight Heroes, Five Righteous Heroes and the Cases of Judge Pao").

References

Bauman, Richard, ed. (1994). *Folklore, Cultural Performances, and Popular Entertainments: A Communications-Centered Handbook.* New York: Oxford University Press.

Bauman, Richard. (1986). *Story, Performance, and Event.* Cambridge Studies in Oral and Literature Culture 10. Cambridge: Cambridge University Press.

——. (1977). *Verbal Art as Performance.* Prospect Heights, Illinois: Waveland Press.

Bosco, Joseph (1994). "The Emergence of a Tawanese Popular Culture," in Murray A. Rubinstein, ed., *The Other Taiwan: 1945 to the Present.* Armonk: M. E. Sharpe. Pp. 392–403. Originally published in *American Journal of Chinese Studies,* 1: 51–64.

Ch'en, Chengchih (1991). *Changchung Kungming: T'aiwan Te Chu'ant'ung Ouhsi* ("Glory in the hands: The traditional Taiwan puppet theater"). Taichung: Taiwan Provincial Government.

Foley, John Miles (1995). *The Singer of Tales as Performance.* Bloomington: University of Indiana Press.

Government Information Office (1997). *The Republic of China Yearbook 1997.* Taipei: The Government Information Office.

Humphrey, Jo (1980). "The Fujian Hand Puppets in New York." *CHINOPERL Papers* No. 9.

Kapchan, Deborah A. (1995). "Performance." *Journal of American Folklore.* Pp. 479–508.

Li, Ho (1993). *T'aiyu te Chihhui* ("The Wisdom of Taiwanese"). Taipei: Tao T'ian Publisher.

Liu, Huanyueh (1990). *Fenghua Chuehtai Changchungi: T'aiwan Ti Putaihsi* ("The unsurpassed elegance and intellectual brilliance of the hands' art: Taiwan's hand puppet theater"). Taipei: Taiyuan Publisher.

Rubinstein, Murray A., ed. (1994). *The Other Taiwan: 1945 to the Present.* Armonk: M. E. Sharpe.

Wachman, Alan M. (1994). "Competeing Identities in Taiwan," in Murray A. Rubinstein, ed., *The Other Taiwan: 1945 to the Present.* Armonk: M. E. Sharpe. Pp. 17–80.

Periodicals

Sinorama Magazine, Chinese–English bilingual monthly. Vol. 23, No. 2, February 1998.

Central Daily News, Overseas Edition, in Chinese. March 7, 1997.

Central Daily News, Overseas Edition, in Chinese. July 29, 1998.

World Daily News, in Chinese. July 2, 1998.

Chapter 5

The Quest for Difference versus the Wish to Assimilate: Taiwan's Aborigines and their Struggle for Cultural Survival in Times of Multiculturalism[1]

Michael Rudolph

The claim for multiculturalism in Taiwan's political and cultural sphere since the early 1990s affects Taiwan's Austronesian population and the cultures of these peoples in various ways.[2] It can be observed that particular segments of Aboriginal society, which may differ in social strata as well as in ethnic backgrounds, often have completely different and even mutually excluding views on the question of which parts of a particular culture should be preserved, revitalized, renewed or omitted—a divergence that has become even more evident in the course of a growing sense of "culturalism" in ethnic elites[3] during the last couple of years: the need to demonstrate cultural and ethnic particularity felt by the latter—a requirement that evolves from the "discourse of difference"—often forms a sharp contrast to the desire of ordinary people to assimilate to social norms. Especially those strategies which aim at a deconstruction and subversion of authoritarian structures of dominance and Han-centered thinking are mostly met with ignorance and refusal. Hence, despite all its positive implications for Taiwanese society, multiculturism also fosters new contradictions and tensions that challenge the process of further democratization.

Tian Guishi's Homepage "The Facial Tattoo of Tayal"[4]

Looking at Tian Guishi's internet homepage "The Facial Tattoo of Tayal," users worldwide are confronted with impressive and exotic pictures: photographs of old men and woman with greenish-blue tattoos on the chin and forehead, in the case of the men rather decently done, but somewhat more shocking in the case of the women, whose lower part of the face is sometimes totally covered by the tattoos. In one of the attached Chinese language articles Tian—who is himself a member of the Taroko, one of the subgroups of the Atayal—explains the myth of the origin of the custom, in another article he explains the qualifications men and women needed to demonstrate in order to receive the tattoo and acquire the right to marry: men had to prove their skills in hunting and in battle, while woman were expected to have high skills in weaving. Males who were successful several times in headhunting were authorized to add special tattoos to their breast, feet and forehead. Among the stories reported, is that of 90-year-old Biyang Lahang, who had observed the bloody scenes of headhunting with her own eyes and who could describe the way the heads were treated after headhunting.

Tian explains the reasons for his decision to engage in cultural preservation work as follows: he wanted to protect the dying culture of his people from further misunderstandings and humiliations. Many Aborigines and Han of Taiwan did not have any knowledge of their own history: though they had learnt how Sun Zhongshan overthrew the Qing dynasty, they had never heard of the anti-Japanese martyr Mona Ludao.[5] Tian, then, tells the story of his son who had been ridiculed because of the tattoos of his relatives, who looked like members of the Yakusa to his schoolmates. However, during the harvest festival in 1993 Tian was surprised to observe the Han carrying the knives of savages, dressed like Atayal and sitting on the seats of the Atayal elders, and thought to himself that these seats actually belonged to the elders. Another motivation for his work was the rapid disappearance of tattooed people: in 1993, when he was the people's representative in Xiulin township, there were still 82 of the tattooed people living, but in 1996 he found only 34 of them left.

In all, the biggest problem Tian faces is not the contest with time, but the reluctance of his own people to cooperate with him in his work of cultural preservation. For instance, he has even been hounded by the dogs belonging to the tribe, while pursuing his documentation work. Nevertheless, by 1997 he had succeeded in filming more than 100 of the tattooed faces of tribal members and recording the life histories of these individuals. One of the few moments of encouragement during his often frustrating and fatiguing work were the exhortations of one of the accompanying Han journalists, who expressed the hope that "the lost tattooing-culture of the Atayal would some day return to its tribes, so that following generations would come to know the glory of the past of their people."[6]

The homepage just cited is a good example of the way that Aboriginal culture today is represented by members of both the Aboriginal and the Taiwanization movements. We find references to the high value and the particularity of the dying Atayal culture, to the cultural practice of headhunting and its connected customs, to the devaluation of these customs by the Han and to the "rehabilitation." This has only taken place recently and is, on the one hand, due to the awakening of the Aborigines who have realized the value of their cultures through the fetishisms of the Han. But it is, on the other hand, also due to the attention that the Aborigines receive from Han intellectuals and Taiwan's media, who are increasingly inclined to recognize and acknowledge the Aborigines' value in providing testimonies of a non-Chinese past and as themselves being representatives of alternative value systems. But there is still something else to be learnt from the homepage: while the protection of Aboriginal culture pursued by Aboriginal elites is obviously very much supported and encouraged by the Han, ordinary Aborigines seem to have problems in identifying with the cultural perspectives and value orientations of their elites.

Acceptance of the Cultural Perspectives of Elites within Aboriginal Society

The results of field research conducted in villages of the Taroko and the Paiwan from 1994 to 1996 with the aim of evaluating the acceptance of the Aboriginal movement serve as a confirmation of this picture.[7] In the case of the Taroko, few people regarded the tattoos as an expression of "culture"; in most cases, these signs of "savageness," and those who still wore them, were hidden as far as possible. People wished even less to talk about headhunting. Instead, I was often told the story of Ji Oang, the Taroko woman who brought Christianity to the Taroko under the Japanese, and the plight and the suffering of missonaries such as Wilang Takao, who was said to have endured severe punishment for evangelizing Aborigines in Japanese times. Despite the cruelty of the Japanese, most people said that they would not blame them, because after all the Japanese liberated the Taroko from headhunting even before the arrival of Christianity. The name of Mona Ludao was, in fact, really largely unknown, only a few older people knew that he must have been a Dekedaya or Bleibao (another subgroup of the Atayal) and not a Taroko. Some of the younger people knew the name of Mona Ludao by having read a *bentu* (a comic book) with the title "The Wushe incident."[8]

As with the attitudes concerning the headhunting past of the Taroko, the conceptions of origin often formed a contrast to the convictions of the elites: only a few villagers were inclined to regard themselves as "Austronesians," that is, as members of peoples who were totally different from the Han. They had already become used to the belief that they were of common origin and descent as the Han people (including the affiliation to a 5000-year-old mainland culture), just as the KMT educations they had received had assured them for decades, in spite of the daily allusions to their backwardness, testified through their "dialect speaking" and their differing living and housing styles. They had also internalized the view of history proclaimed by the KMT until the early 1990s, according to which some day in the future the mainland would be recovered and ruled again. In some cases, I was told how "one" (that is, the Chinese) had been mistreated by the Japanese during the "eight-year anti-Japanese war," and that it was forty five years since "one" (that is, the ROC) had come to Taiwan. In contrast, the political situation of the Aborigines in Taiwan was not very well known: very few people knew that the only central government institution for minorities was dedicated to Tibetans and Mongolians and that there was no similar institution for Taiwan's Aborigines: one of the improprieties the elites were fighting against.

From this perspective, it seemed totally useless and even against one's own interest to reinstate traditional first and family-names, as had been allowed by the government in January 1995 after many years of engagement by the elites. Many said that there were too many different names in their families already; others believed that a traditional name would make them indistinguishable from other Tarokos with the same name. And, last but not least, a reinstating of traditional names would only make sense if everybody in the family consented, which seemed very unlikely under the existing conditions.[9]

Likewise, the people could not see a crisis of their mother language in the same sense as this was perceived by the elites: the Taroko language was widely used, but many people also believed that they could live without it (English was believed to be more important).[10] The same was thought of the durable or eternal possession of mountain reservation land: it was of equal importance to make investments with the earnings from it (in many occasions after selling it illegally to the Han), so that one could afford an estate or a home in the cities. Autonomous zones did not seem very attractive from this point of view; it was even suspected that this was only a means to get Aborigines "locked up in a cage so that you could look at them like monkeys in the zoo."

The Taroko villagers were not the only ones who regarded the activities of the elites, to revitalize and protect culture, with suspicion. In the Paiwan village where I stayed, I realized that the skepticism against official rehabilitation of traditional front and family-names was especially strong. Due to the rudimentary subsistence of certain structures of the former nobility and class society (which was partially a consequence of the government instrumentalization of people with former nobility status), non-noble members of this society, naturally, regarded the possibility of name rehabilitation with very mixed and ambivalent feelings: an official rehabilitation of one's status-revealing front and family-name would inevitably cause a fall back into one's former subordinate, inferior status.[11] Thus, often, they even even refused to tell me their "bad-sounding" Paiwan names. In contrast, the former "nobles" with their "nice-sounding" names tried to make use of the favorableness of the situation and emphasized the superiority of their class in *bentu* publications, schoolbook-materials and in newly established "culture protection committees."

A Methodological Excursus

In a discussion on the construction of the past in the South Pacific, Roger M. Keesing (1989) describes the origins and functions of modern myths.[12] According to his findings, many discourses of cultural identity in postcolonial Melanesia and Polynesia have developed in constant interaction with Western ideologies. As he shows, the categories of the dominators were extensively internalized, not only because the discourse of domination created the objective conditions in which struggles must be fought, but also because it defined the semiology in which claims to power must be expressed. Nevertheless, Western ideology was often not taken over directly; instead, parts of indigenous culture that were believed to differ most strikingly from the dominant culture were selected and confronted with the former in a dialectical way. Common examples are idealizations of "sharing," "communal life" and "unity with land and nature." However, many of these idealizations of the precolonial past, which were formulated by educated, careerist elites, were very similar to idealizations of primitivity, wisdom and reverence for ecology put forward by critiques of modern technology and progress. Further, Keesing mentions that the identity-endowing idealizations of the past were often based on anthropological concepts (it thus seems ironic that it is precisely the anthropologists who are frequently accused of "exploiting" indigenous cultures). But, as even these "real" pasts can only reflect partial realities—because they include and transport the essentialisms, romanticizations, mystifications and fetishisms of the anthropologists, or because they rely on interpretations of former ruling elites—it is not so important to raise the question of the relationship between "authentic" and "inauthentic" culture. What matters more is the question of how the legacy of those "real" pasts influences the present, for instance, by way of certain power structures. For this reason, Keesing demands:

> A critical scepticism with regard to pasts and power, and a critical deconstruction of conceptualizations of "a culture" that hide and neutralize subaltern voices and perspectives should, I think, dialectically confront idealizations of the past.[13]

In the section that follows, I will show that such skepticism is also necessary with respect to the reconstructions of the past undertaken by Aboriginal elites in Taiwan: the contradictions between the elites and people I mentioned earlier often have their origins here. However, in Taiwan, the mutually mirroring levels of dominators and dominated seem to be even more complex. In their discourses, which often draw heavily on Western theories, Aboriginal elites not only relate to their Taiwanese dominators, but to mainlanders and Taiwanese simultaneously, who themselves face each other in a postcolonial relationship; moreover, people in Taiwan are also forced to cope and to deal with threats of incorporation from mainland China. But today, these three counter-hegemonic discourses are not clearly separated anymore: they mutually fertilize and give wings to each other, often by utilizing western theories and concepts, but also by excluding the less educated, who are not able to follow the rapid changing meta-discussions or who just do not see any advantages in certain ways of representation. While some of the members of the Aboriginal movement had visualized this incongruity already by the end of the 1980s—this caused them to proclaim the "Return to the tribes movement"—multiculturalism had even enhanced these contradictions. For a better understanding of the interrelationship between multiculturalism, the role of Aborigines and the commitment of Aboriginal elites I discuss background and the development of multiculturalism in Taiwan.

Multicultural Taiwan

Simultaneous with the democratization process, which has been going on since the lifting of martial law in 1987, we also observe a steady revival of ethnic and cultural identities in Taiwan.[14] The homogenization and amalgamation of Taiwanese society as it had been pursued by the KMT previously—embodied in slogans like "Children of the Yellow Emperor"—seems to belong to the past.[15] With the increasing efforts of the Taiwanese to point out their differences from the mainlanders as well as from mainland China with respect to culture, history and consciousness, the former "question of provincial descent" has developed into an "ethnic question."[16] It was at this time that claims for recognition of the multiculturality of Taiwanese society and the implementation of multicultural politics became louder every day.[17] By the beginning of the 1990s, not only governmental institutions like the *Council of Cultural Planning*, but also politicians from the opposition party referred more and more often to Taiwan's society as a "multicultural society." This pointed to a re-introduction of cultural-ethnic differentiation into a society which had earlier to a large extent already been functionally differentiated.[18] Almost imperceptibly, the postulate of the monocultural, homogenous society had been replaced by the "discourse of difference."

This new self-description "multicultural" not only added a new dimension to Taiwan's democratization discourse, but also caused an inherent dilemma of democratic systems—that is, the precarious dialectic of "universalism" and "particularism"—to become even more salient.[19] Now it had to be asked to what extent the claim of equal rights, equal respect and non-discrimination could be satisfied by a politics of "recognition of universal human dignity," or whether cultural difference should be recognized to a much larger degree than before in order to give non-mainstream members of the "life-(or fate)-community" on Taiwan the feeling of a more respected existence.[20] Under such circumstances, their "cultural difference" would be taken as the basis for a differential practice. They would be guaranteed certain rights and authorities that did not apply to other Taiwanese, and—as multiculturalism in its deepest sense also suggested—attention would be paid to those interests which aimed at the cultural survival of a group and the generating of further members.[21]

Taiwanese Multiculturalism

By looking at Aboriginal politics, we can see very clearly that some steps in the direction just mentioned really have been made. Such a development seems particularly astonishing, as in the past all administration measures regarding Aborigines were handled as "temporary regulations" which would soon become unnecessary.[22] Some initial self-criticism of previous Aboriginal policy and its results was put forward in the

"Program for Mountain Society Development" set up by the provincial government in 1988. In the same year, the government announced the setting up of a five-year plan to improve Aboriginal education. The plan was supposed to contain the following aims: the promotion of contact and communication of mountain-society with the main society; the promotion of marketability; the preservation and promotion of Aboriginal languages and cultures to build up self-dignity and self-respect; the promotion of talented people to develop the capability for autonomy. The five-year plan was finished in 1992 and put into force in 1993, the International "Year of Indigenous People." In that very year Guo Weifan, Minister of Education, and Wu Boxiong, Minister of the Interior, openly admitted mistakes in former education policies and promised the implementation of classes in vernacular languages and local knowledge by 1996. In 1994, the government proclaimed a plan for the implementation of elementary- and junior-school education in preferential zones, which was supposed to meet the disparities in education between countryside and cities by means of "active reverse discrimination."[23]

But the "recognition of difference" was not limited to the field of education: important concessions have also been made in general policy, for instance, concerning the recognition of the self-chosen name of the Aborigines, "*Yuanzhumin*," in 1994, the right for rehabilitation of traditional front and family-names in 1995, and the establishment of "Aboriginal Affairs Committees" not only in the two metropoles, Taibei and Gaoxiong, but by late 1996 also on the central level, with representatives of all ten different ethnic groups, including the Peipohuan (*pingpuzu*), which had reappeared in 1990. After 1991 the government also gave increasing attention to Aboriginal communities in the course of its efforts toward "community reconstruction."[24] Every ethnic group was now encouraged to search for its own cultural particularities.[25]

The official change toward multiculturalism also caused a change of the government's attitude toward the oppositional Aboriginal elite. In the course of the cultural reconstruction of Aboriginal society, their members were increasingly integrated and engaged into projects initiated by central government institutions. The Ministry of Education and the Council of Cultural Planning now became frequent dispensers of jobs. Since 1992, the teachers' college in Hualian has organized regular classes for Aboriginal teachers as well as for Aboriginal students of teachers' colleges who were to teach in Aboriginal schools, to improve their teaching ability in themes related to Aboriginal culture. Furthermore, teachers have been encouraged to participate in the work of creating Aboriginal teaching materials. The central government, thereby, joined the efforts of the opposition, who had started to engage Aboriginal elites in the education sector as early as 1990 (just about the time when the opposition also started to organize homeland and vernacular education).[26] The development depicted here also led to an increasing amalgamation of the two originally antagonistic and mutually-despising wings of the Aboriginal elites, that is, the oppositional and the KMT-loyal, political elite.[27]

The Background and Functions of the Multiculturalism of the 1990s and the Role of Aborigines

If we ask for further reasons for the development toward multiculturalism and the role played by the Aborigines in this process, we find some hints in the writings of Walisi Yougan.[28] In a critical discussion on Aboriginal vernacular education, the young Atayal writer argues that the phenomenon of multiculturalism in Taiwan has to be seen in close relationship with the efforts of "Taiwanization," the "deconstruction of the authoritarian system," the "discovery of Taiwan," the "return to the homeland" and the "search for Taiwanese subjectivity." As Walisi points out, even the initiative to implement vernacular language classes was not so much due to the latent ethnic consciousness of the Aborigines, but to the endeavor of local DFP and KMT governments to show their willingness and fervor for Taiwanization. The Paiwan and political sciences scholar Gao Deyi points to some additional grounds for the implementation of multicultural politics.[29] In his article, "The Development of Ethnic Relations in a Pluralistic Entity and Aboriginal Politics in Taiwan," he names the functions of an adequate Aboriginal policy: it can serve the realization of the equality of nationalities as provided in the constitution; it further helps in strengthening Aborigines' loyalty toward the government, assures the healthy development of Aboriginal society, strengthens cultural protection and fertilizes national culture, lifts the international image and enhances the peaceful competition with the mainland. The ethnologist, Wu Tiantai, director of the Aborigines Education Research Center at the Teachers College in Hualian until 1996, explains the necessity for the implementation of multicultural education as follows: The lack of respect toward the coexisting ethnic groups that had been expressed through sinicizing cultural policies caused their members to develop that kind of social stigma and feeling of inferiority that ethnologists like Xie Shizhong and Xu Muzhu described as constituting the main source of adaptation problems and which had a negative impact on ethnic interaction. Wu emphasizes that a multicultural people must not necessarily have a common ancestor to develop the imagination of belonging to the same "fate community." In the same article, Wu points out that by learning more about Aboriginal culture, students can exercise their ability for analytical thinking. In this way they learn how to catch up with the needs of modern sociey. This means that multicultural education not only aims at the improvement of Aboriginal education in the schools, but also helps to improve the education of the whole people.[30]

All this shows that members of the Aboriginal elite are very much aware of their value in Taiwan's society today. They know about the potential that Aborigines are believed to have in the area of the construction of Taiwanese identity, directed inward as well as toward the outside (for instance, toward the United Nations or investment partners from the South Pacific);[31] they have recognized their usefulness in being instrumentalized against the conservative wing of the KMT or against the incorporation efforts of the People's Republic of China. And they are also aware of their significance for the fertilization of Taiwan's cultural climate.

Striving for Authenticity

It was this new attention which the Aborigines and their cultures received from growing segments within Taiwanese society (political opposition, Taiwanization-oriented circles within the central government, ethnologists, human rights organizations and environmental protection groups), that caused Aboriginal elites to develop a new kind of self-confidence and self-consciousness. More and more people within the elites now realized the importance of the protection and, if necessary, the revitalization of Aboriginal culture and ethnicity. The question of "authenticity" at this time also became increasingly significant for the elites in the process of forming alliances.[32] This can be seen by the growing support and commitment of the former KMT-loyal Aboriginal elites with regard to legal recognition of the Aborigines' ethnonym and status, for an Aborigines' basic law, for the rehabilitation of traditional names, for the implementation of Aboriginal institutions on the central level, as well as for autonomous zones—matters that so far had only been fought for by the oppositional elites whose members mostly originated from church and opposition circles or from campus student organizations.[33] As for the work of preservation, protection and revitalization, great hopes were now placed in those Aboriginal elites who went back to the tribes as social activists, teachers or ministers to "save what still could be saved." Large expectations were also projected on the twelve Aborigines who were instructed in the area of documentary film by Public TV, the channel from the central information bureau, and who from 1994 on traveled through Aboriginal villages to record traditional rituals and festivals. Several private filming also began to engage Aborigines as filmmakers.

The growing degree of interaction between people and elites caused latent contradictions and differences in cultural perspectives to become more salient. What can be named in this context is the failure of the renovation and rehabitation work of Old Haocha, or the tensions that developed in the course of the protest activities against the building of the Majia water reservoir because parts of the Rukai population of the village, threatened by inundation, were not opposed to the idea of being resettled to the infrastructurally better off plains. A good example of the contradictions between the elites' "striving for authenticity" and the peoples' "understanding of cultural practice" is a situation I experienced when attending the combined harvest and fishing-festival of the Amis in Qimei.[34] While the people were willing to cooperate with Han film director, Yu Kanping, in order to raise the glory of the tribe—Qimei was known for the most "authentic" festivals and the best preserved year-rank system within the Amis—they reacted quite angrily when the filming elites decided that intruders from the outside should not be tolerated during the filming activities because this was against the rules of the ancestors. In the eyes of the commoners the integration of a foreigner into the dances and into one of the central initiation rituals (which became necessary because of the lack of real Aborigines in one of the year-ranks) only helped to make the very exhausting ceremony more vivid and exciting; after all, it was not believed to be of any hindrance to the honor of the tribe. That they were wrong with regard to this last point was soon proved by the reactions from some of the Han spectators, who openly expressed their indignation at any intrusion into the still "intact" year-rank system of Qimei (Han spectators in Qimei at that time mostly originated from the intellectual "scene"). A similar contradiction is described by Xie Shizhong in his article "Tourism, the Shaping of Tradition, and Ethnicity."[35] Xie focuses on those Atayal from Wulai who work in the tourism sector; in order to adapt their cultural productions for the amusement of the Han tourists and to meet their expectations for the exotic, they do not object to the synthesis of Atayal culture with foreign elements. Interestingly enough, they do not regard this self-made hybridized culture as a false culture, but identify themselves with it. Local intellectuals such as teachers or ministers, on the contrary, reject this commodified culture because it does not match the "authentic" Atayal culture displayed in the museum, which mainly consists of anthropological materials.

These observations suggest that contradictions between elites and people with regard to cultural praxis may develop because different segments of Aboriginal society attach themselves to different value-orientations within Han society: as the work of culture preservation and revitalization pursued by Aboriginal elites is frequently morally and financially supported by Taiwanization circles, environmental protection groups and the like. Aboriginal elites also often identify or at least sympathize with these worldviews; in contrast, commoners feel much more attracted by the value orientations of a consumption-oriented Han middle class.

The Quest for "*Yuanzhumin*-Subjectivity": The "Re-emergence of Headhunting"

However, as already mentioned at the beginning of this chapter, not only does the "striving for authenticity" of the elites sometimes lead to tensions and contradictions with the perceptions of commoners, but also the way the image and status of the *Yuanzhumin* are reconstructed and described today. So, where do the representations undertaken by the elites derive from that they are so different from the expectations of the ordinary people? I would now like to return to my introductory example of elites referring to tattooing and headhunting culture, because here we can see the interaction of certain value orientations.

When I first started to explore the situation of Taiwan Aborigines and the related social problems in 1987/88, besides child prostitution,[36] the cases of Tang Yingshen and Dongpu and the "return our land" debate, yet another topic attracted great public attention: the discussion on the negative impacts of the "Wu Feng story," which until 1988 was still part of the history teaching-material in primary schools and for most Taiwanese was their first and sometimes only occasion of any kind of contact with Aborigines.[37] It was the anthropologist Chen Qinan, later vice-head of the Council of Cultural Planning, who in 1980 first expressed doubts about the verifiability as well as the adequateness of the story reprinted in schoolbooks. By this he initiated a hot debate, in which not only anthropologists but also members of the opposition, the Presbyterian Church of Taiwan (PCT) and the Alliance of Taiwanese Aborigines (ATA) were to take part. Most severely criticized was the representation of Aborigines as "raw, wild and morally rotten," as was suggested in the "legend" through its emphasis on their indulging in headhunting and the mean murder of the noble Confucian Wu Feng. As protests did not cease, in 1988 the story was taken out of the schoolbooks; the same year, the Wu Feng memorial statue in Jiayi was torn down and smashed by a group of Aboriginal activists. It now was regarded as more or less politically incorrect to mention headhunting in relation to Aborigines, and even anthropologists seldom referred to it.[38]

After these impressions, it was quite confusing for me to be confronted with "headhunting" again in April 1994 when I attended the "First Aboriginal Culture Congress," where not only Aboriginal activists and anthropologists, but also politicians participated. The day the congress began, a group of ten Aboriginal activists in traditional costumes suddenly marched on to the stage and openly announced the "cultural headhunting raid proclamation."[39] In general, this was a catalog of demands in which Aborigines requested to be taken more seriously with respect to their sovreignty. The participating cadres were asked to intensify their cooperative efforts regarding name correction[40] as well as the implementation of Aborigines' institutions and autonomous zones; the anthropologists who had been the main planners of the congress were blamed not only for wasting too much time on academic questions, but also for their Han-centered worldview, demonstrated by the under-representation of Aborigines at the congress, . . . and further, the whole agenda of the congress, that they said, should be changed to reflect the Aborigines' perspectives.

But this allusion to headhunting was by no means the only reference to a theme which had originally been banned from public discussion some years before. In the following months while I was doing my field research in Taiwan, topics like headhunting and the possibility of the continuity of the *gaya*—the laws of the Atayal—again and again came to my ears frequently. For instance, I heard speculation about the mysterious death of Duo Ao, who didn't simply die in a car accident but had tried to act in accordance with the *gaya*; or I was told stories about the last headhunting incidents on the east coast in the 1950s. Also, in Aborigine literature of the 1990s there was an increasing tendency for allusions to headhunting, for instance in a book by Walisi Yougan, "Drawing the Savages Knife."[41] While it had already become evident in commentaries during the Wu Feng debate that there was a willingness on the part of Aboriginal activists to falsify not just the "savageness and meanness of the Aborigines," but to relativize it as a style of representation inherent to the Confucian value-system.[42] Walisi now appealed to the Aboriginal elites to stand up against the "enslavement" by the state: in order to be successful, one should be "equipped with the will of the hunter who takes revenge for former humiliations," otherwise one would be thoroughly "civilized" and corupted by "civilized society" (which had built its civilization on exploitation etc.).

From the perspective of colonial and postcolonial discourse—that had been adapted by intermediation through the anthropologists as early as the 1980s—this way of proceeding by the elites was to a certain degree understandable: in order to reach thorough emancipation one had to liberate oneself from the negative self-image that the dominators had forced upon the dominated. Fanon suggested violence as the way to liberty—violence as an equalization for the violence one had endured in receiving the negative self-image.[43] But how did it come about that the negative image itself was now taken up again by the Aboriginal elites, despite the well-testified repugnance of the ordinary people toward it?

The Reasons for the Change of Attitude toward Headhunting

First possible answers on how this change could be explained were received through my acquaintance with the Han caricaturist, Qiu Ruolong, and his book *The Wushe Incident*, a work greatly revered by Aboriginal elites and published by Qiu as a comic after several years of stay with the Atayal of Wushe.[44] Besides being an extensive biography of Mona Ludao and a depiction of his role in the fight against the oppression of the Japanese intruders, the book also contains a thorough reevaluation of the moral systems and value systems of the Atayal. Qiu, here, works with representations of the tattooing, and the headhunting culture of the Atayal, which are as fascinating as they are shocking. In a commentary on the book four years after its publication on the occasion of the National Festival of Culture and Arts organized by the *Council of Cultural Planning*, Qiu offers the following explanations for his motivation in doing research on the Wushe incident and the tattooing-culture of the Atayal:[45] "It was of great importance to him to mediate a significant and great truth which was hidden behind the tattoos of the old people. Because of the dying away of the old tattooed people, this practice would very soon not only vanish, but might also be misunderstood as an expression of savagery. The value system of an entire people would then be lost and have been substituted with modern science civilization." As for the tattooing culture, Qiu explains:

> With a face naked like that of an ape, you wouldn't belong to the human race. Only courageous men and capable women were allowed to produce descendants. Under such rigid conditions, men who didn't capture heads and those who where not courageous enough as well as those who were captured themselves were naturally eliminated, the same as the lazy and

> the dull-witted women. Thus, it was a kind of "eugenics" or "qualification certificate." According to the aesthetic conception of the Atayal this was considered as "beautiful". Through it, one's own people could be distinguished from the enemies, and after death it was this sign by which the ancestors would recognize you and allow you to enter "paradise."

The facial tattoo of the Atayal, Qiu, continues, must be considered as an explanatation for why these people have been able to survive for such a long time. Their tattoo showed their nature-revering spirit. In times of hunting and slashburning, the earth could only feed a limited number of people: that's why the Atayal developed a culture "adaption to nature without changing it" and "unity with the natural ecological equilibrum." Qiu then concludes with the words:

> When you look at the destruction caused by modern civilization in Taiwan, you ask yourself how long mankind can still live here. Thus, the old people with facial tattoos are not only a national treasure witnessing old culture. They are outstanding personalities in which abilities, virtues, art, philosophy and practice are concentrated. . . .

The positive reevaluation of Atayal culture undertaken by Qiu can, nevertheless, only partially explain the change of attitude of the Aboriginal elites.[46] After all, Qiu's interpretation is probably much more an expression of change that was already happening. Thus, a question that might lead us further is why people in Han society should actually be interested in such a reinterpretation of the value-systems of Aborigines.

References that hint at an historical interest in headhunting can be found in newspaper commentaries on the occasion of the sixty-fifth commemoration day of the Wushe incident in 1995 (the first big commemoration festival was held in 1990). Several authors here discuss the question of whether the Wushe incident was really an expression of anti-Japanese opposition by China-loyal Aborigines, as it had been described by the KMT, or whether it expressed the desire of the more or less Japanized Atayal to revitalize headhunting after this practice had been prohibited in 1914. What would confirm the latter are the headhunting rituals held directly after the incident.[47]

That there was more than a pure historical interest in the differing value-systems that were manifested through headhunting is testified by the commentaries that were published by Han intellectuals in *Taiwan Indigenous Voice Bimonthly* directly after the Aboriginal culture congress. In her analysis of the relationship between anthropologists and Aborigines, the journalist, Chen Shaoru, writes with satisfaction that at the congress, the Aborigines for the first time expressed their subjectivity in front of the Han-cadres and ethnologists by doubting the use and the functions of this gathering and by showing that they were no longer willing to be "discussed" and "researched" objects of the ethnologists.[48] And in an article entitled "The culture headhunting proclamation is the beginning of a dialogue between Han and Aborigines," the Han and producer of documentary films, Jiang Guanming, points out that the Aborigines should have their own strategies and discourses in order to secure their space for existence and to construct their cultural subjectivity and dignity.[49] They should not make themselves depend on the decisions and interpretations of the government or the anthropologists. Furthermore, Jiang emphasizes the influence of the cultural interpretations of the Aborigines and the tension generated by this for the development of the Taiwan discourse. He says that it "is the question of Taiwanese subjectivity that is touched on here, . . . even to a much larger extent than in the home-literature debate or in the modern literature movement."

Taiwanese Subjectivity and Aborigines Subjectivities: The Search for New Paths for the Deconstruction and Subversion of Han-hegemony and Han-centrism (or: "De-Nobling of the Noble Confucian by Ennobling the Savage")

Here, we finally see the significance of headhunting allegories within the cultural and political context of Taiwan in the 1990s: what first seems absurd, serves the manifestation of "subjectivity." Nevertheless, the manifestation of "Aborigines subjectivity" that has been clearly demonstrated by the emancipation from the ethnologists, and also the "Taiwanese subjectvity." The historian Chen Zhaoying comments as follows:[50]

> Until the beginning of the 90s it became clear that the concept of "Taiwan consiousness" was too vague, thus it was almost totally substituted by the concept of "Taiwanese subjectivity."

Comparing the commentaries of a couple of different authors, Chen, then, analyzes six contrasted pairs, which are obviously included within the concept of "Taiwanese subjectivity," that is: China/Taiwan, center/periphery, dominator/people, from the outside/homeland, non-independent/independent, without subjectivity (colonized)/subjectivity. For Chen this means that on the one hand one seemed to set up the equation "China = Center = dominator = from the outside = non-independent = without subjectivity." And on the other hand, there is this equation "Taiwan = periphery = people = homeland = independent = subjectivity." From these equations it could be concluded that the realization of subjectivity will only be possible by separation from China. But, Chen warns,

> Suppose that Taiwan [in the name of subjectivity] really succeeded in detaching itself from the domination of the center China if then there existed further domination in its interior—no matter whether between members of different provinces, classes, ethnic groups or genders—then the legitimation for detaching oneself from China would suffer damage, and the construction of subjectivity would also be totally impossible.

Other scholars had also recognized the danger that Chen describes here. That is why they pleaded for a radical abolition of Han-centered and Han-chauvinist thinking. At the inaugural symposium of the Wusanlian Foundation, the ethnologist Xu Muzhu made the following remarks:[51]

> Though we often criticize the domination manners of the Han from the mainland, we ourselves frequently approach the "savages" of Taiwan with the attitude of the Han from Taiwan. . . . When interpreting Taiwanese history we should try not to assume a Han-centred attitude. In the historical conception of the so-called "Taiwanese subjectivity" the viewpoints of all different ethnic groups in Taiwan's history and prehistory must fuse.

From this perspective, the manifestation of any kind of minority-subjectivity not only had to be tolerated, but was even absolutely necessary if one wanted to convince others and oneself about the sincerity and the maturity of "Taiwanese subjectivity."

Indeed, one had been waiting for initiatives from the side of the Aborigines for quite a while. This is demonstrated by the remarks of Sun Dachuan, the chief editor of *Taiwan Indigenous Voice Bimonthly*, made in an article in the first edition of this magazine:[52]

> Learning from the experiences of ethnic minorities in the Third world, some of the scholars who observed the movement of Taiwanese Aborigines began to critically analyze the situation and the literary activities of the Aborigines. (. . .) In general, Aboriginal discourse in the Third world tries to analyze problems from the question of "power." That's why the scholars interpret the whole movement of Taiwanese Aborigines as an activity directed against violence and oppression. (. . .) At the same time, they also realized that the Aboriginal movement—in contrast to the movement of the Minnan and the Hakka—was not only

> directed against the authoritarian regime, which hides behind political and economical suppression, but also against the superior "main culture," which exerts cultural domination. Thus—the scholars say—the Aborigines should attach some importance to the manifestation of their independence and their subjectivity in order not to get caught within the logic- and thought-system of Han culture.

And quoting the historian and social critics, Fu Dawei, from Qinghua University, Sun continues:

> If the Aborigines want to maintain a certain independence and subjectivity in their opposition against suppression, they have to show incessantly and actively strategies and initiatives in the future.

The kind of fertilization the Han expected from the "initiatives" of the Aborigines can be seen in the broadly discussed article by Fu Dawei, "Hunters of Chinese Characters in the Forest of Han Rascals," from 1993. Fu, here, emphasizes the potential of subversion within Aborigines' literature:[53]

> Crucial are perhaps those effects of irony, challenge, subversion and seduction that this writing culture can generate when it succeeds to enter, settle and develop within the writing culture of the *bailang* [Han rascals].[54] (. . .) As for the *politics of language* in Taiwan, the delicate and complicated relationship between the Beijing-Mandarin, the purposely neglected Taiwan-Mandarin, the language of the Holo which becomes the mainstream and the language of the Hakka . . . wouldn't it be possible that the latent explosive potential which is inherent to the grammatical displacements and subversions undertaken by the "character-hunter" of the Atayal can evoke a new politics, a new history and even a new geography within the language and the scripture of the *bailang*?

For Daiwei, like Chen Zhaoying and Xu Muzhu, expresses the anxiety that in the course of the redetermination of Taiwanese culture and realignment of power and resources one single group—that is, the Holo—would again gain supremacy. Then, there would be a high risk of Han-centered and Han-chauvinist value orientations continuing to exist unaltered, and different groups in society might again be culturally, politically and economically suppressed or discriminated against because of their cultural or physical differences. The liberalization project of Taiwanese society then was bound to fail, because within Taiwan suppression would still prevail, and Taiwan would lose legitimacy for the claim that because of its different sociocultural conditions and predispositions it had to walk a different path than mainland China.

"Multiculturalism" in Taiwan, thus, also functions as a "bastion" against the hegemonic tendencies of a Taiwanese nationalism that is rapidly gaining self-consciousness. By pointing out the "intentional neglecting" of Taiwan-Mandarin, Fu implicitly points to the danger of the development of a new kind of cultural essentialism (otherwise why not be satisfied with hybridized Taiwan-Mandarin, which in some way reflects all the different languages of Taiwan?). To counter this newly developing hegemony with other subjectivities seemed to be the right strategy in such a situation. Though their participation in this process was very much desired, neither mainlanders nor Hakka with their *zhongyuan* orientation were suitable to engage in the deconstruction of Han-centrism and Han-chauvinism—the Aborigines seemed to be the only group with the appropriate predispositions.

Some Concluding Remarks

As an economic power that is making increasing efforts to detach itself from China and to obtain political and cultural independence, Taiwan today faces a situation that has "postcolonial" as well as "post-national" traits: the frame in which social processes were organized before—that is, the Chinese national state, in which the political and cultural entities were regarded as identical—is gradually breaking up and disintegrating. Under the claim of bringing about a democratic transformation, limits and rules are newly determined; also newly determined are the possibilities and the opportunities of the players and the distribution of political and cultural resources, social welfare and compensation and subsidizing measures.[55]

However, this process of disintegration and reorientation within Taiwan does not proceed freely and independently, but under the steady impact and influence of another, exterior factor: the threat of a premature intervention or interference from communist China. To the extent that China—conjuring ethno-cultural homogeneity—urges Taiwan to return into the Chinese empire, "heterogeneity," difference" and even "rebellion" receive a new connotation and lose their former negative sense.[56]

It is in this context that we observe two different forms of culturalism today, acting in close symbiosis. On one side there is the culturalism of the government elites (the KMT as well as DPP) that aims at the conquest of old power structures within Taiwan, a demarcation from China and a demonstration of democratic structures vis-à-vis the international community. This kind of culturalism is manifested in "multiculturalist politics," "efforts of community reconstruction" and the "construction of Taiwanese subjectivity," it creates the forum *for*—and needs to be complemented *through*—the culturalism of those who face each other in the process of negotiating social status and political and economical resources. This culturalism frequently manifests itself through "cultural in-scenation": traditions are put on stage (*mise en scène*) testifying to differences, which is the precondition to claim preferential treatment in multiculturalism. I tried to exemplify this by describing the Aboriginal elites' strive for "authenticity" on the one hand and their efforts to accentuate "*Yuanzhumin-subjectivity*" on the other. As for the *Yuanzhumin-authenticity*, this mostly confines itself within the categories desired and wished-for in Taiwanese culturalism: the more convincingly Aboriginal elites succeed in displaying the cultural particularities of the Aborigines the more they can count on the support from the government elites. In this case, the evaluations of anthropologists often serve as standards for what must be considered as different (for instance; typically Austronesian), which claims of minorities are justified and what kind of concessions may be made according to the degree and extent of difference (that is, in language and culture protection, land claims and implementation of legal institutions). As for the "ascertainment of authenticity," ethnologists seldom rely on the ethnographic material of the Japanese or theories of Western scholars; for the establishment of the relevant categories they draw on the principles of international minority politics (as, e.g., in Li Yiyuan, et al. 1983 or Xu Muzhu, et al. 1992). Aboriginal elites have realized very well that the observation of these categories can help to promote their demands more successfully and smoothly.[57]

The conditions under which "*Yuanzhumin-subjectivity*" is constructed are more complicated still. In this case still another discourse is adapted: the postcolonial discourse that had been developed in other Third World countries and that is adapted directly, but through the mediation and interpretation of Taiwan Han scholars. With the help of postcolonial theories they define the kind of treatment that is advantageous for minority-individuals (for instance what kind of self-images should be thrown off) and the cultural strategies that should be adapted to improve a certain situation. In an article on the "Historical status of the *Yuanzhumin*" in *Taiwan Indigenous Voice Bimonthly*, the Qinghua-University anthropologist, Liu Shaohua states:

> The cultural strategy of postcolonial discourse is to develop a new discourse from its experience of border transgression. It transcends the

> political thought models in which the colonizer and the colonized are caught. Only by this can the latter cast off the nightmare of colonization, and culture can begin.[58]

This means that if members of Aboriginal elites today talk about headhunting and stage a "cultural headhunting raid," they surely transcend the provided categories, but they do not take any risks, because besides being a manifestation of authenticity it is just the transcendence of the existing paradigm which is expected from them. Regarded in this light, we may interpret the culturalism of Aboriginal elites in a way similar to how the German ethnologist, Werner Schiffauer, describes the behavior of Turkish migrants in Germany. According to his findings, their culturalism is not so much an effort to bring about demarcations but is an appeal for *solidarity*: people who identify with the same culture produce commonness; for *communication*: people who have produced a common culture can refer and appeal to this commonness; and for *recognition*: people who appeal to a common culture wish for this aspect of their self-understanding to be recognized by the wider society.[59]

Due to this drive for recognition within Han society, one of the dangers that critics of multiculturalism mention can be regarded as minor, that is, that the interpretation of conflicts on ethnic lines would necessarily reduce the willingness to make compromises, as conscience and tradition would then rank before an open compromise-oriented way of proceeding.[60] However, an exception might develop for the question of hunting. From my observation of this, not only ethnic elites but also people in the villages often refer to hunting as something "holy," because it stands for "protecting last surviving traditions" and because it is often still regarded as a source of self-esteem in one's own community (today, hunting is mostly done with traps that have iron teeth!). As long as the support from environmental protection groups and human rights groups is still needed and allocation of resources from the government and the larger society remains as it is, most Aboriginal elites will probably continue to adjust to the values of Han elites. This might change if the larger society's interest in the Aborigines should fade, so that Aboriginal elites find themselves totally dependent on the vote potential within their own people: then the question of hunting might easily be misused for political mobilization. From this perspective, the mobilization of the Taroko population of Hualian against the national park regulations in 1994, must be considered as alarming. One of the reasons this mobilization could be so successful (almost 2000 demonstrators in a "remote area"), was because "anti-governing" elites[61] successfully reminded common people of their obligation to "ethnic solidarity."

What also causes anxieties are the contradictions between the elites and people. Ordinary people often do not see any practical use in the way ethnic elites emphasize differences or even see disadvantages (as in attitudes toward cultural practice, ethnic tourism, language preservation and use of reservation land).[62] But it is also important to pay more attention to the aversion that ordinary people feel when their belonging to a "different people (or nation)," "different civilization" or "different value-system" is too strongly accentuated. Thus, in their memories, large sociopolitical transformation and turmoil are still present. Critics of multiculturalism often point to the dangers inherent in "othering": in the case of social upheaval or quarrels, ethnicity could very easily become a resource again.[63] In times of overall sociopolitical changes (not totally inconceivable in the case of Taiwan) or in case of a throwback to an era of cultural-ethnic dominance of one certain group, the difference that then sticks to one's body could then prove to be fatal (at least for those who cannot escape from their communities).

However, when we look at the situation of the Paiwan, such calamitous prospects are not even necessary to imagine the discomfort that could be caused by an emphasis on the former class-difference—petrified in the traditional front and family-names—for the lower-class members of this society. From this example we can clearly see the limits of multiculturalism (as well as the limits of difference and subjectivity) in Taiwan. Here, it becomes evident that it might be harmful to further democratization to indiscriminately comply with the demands of ethnic elites for the cultural survival of their collectivities—for instance by officially *ordaining* the rehabilitation of names. True, it can be argued that the "right to difference" that is inherent in multiculturalism cannot be limited to individuals: on account of the dialogic character of human existence, this right, in some cases, only makes sense if it is granted collectively, as in the case of language or the rehabilitation of names. But to vest "cultural collectives" or their representatives with rights that enable them to generate further members according to their own perceptions would surely—as the example above suggests—result in discrimination against further, subordinate groups.[64]

Notes

1. Parts of this paper were originally presented at a conference of Taiwanese and American scholars (i.e., The Third Annual Conference on the History and Culture of Taiwan) who gathered at Columbia University in August 1998, to discuss the results of Taiwan's democratization process of the last

eleven years and the cultural perspectives of the country. In order to enhance discussion, this paper contains some undertones critical of Taiwanese multiculturalism. However, this should not be misunderstood as a devaluation of the efforts made toward democratization in Taiwan: after all, democratization as well as multiculturalism (as a necessary "by-product" of democratization in the political and sociocultural context of Taiwan today) brought a lot of improvements to the life situations of people in present-day Taiwan. Nor should this criticism be misunderstood as a devaluation of the Aboriginal movement, whose members have gone through periods of hard struggle before *taiwanization (bentuhua)* became an officially accepted and encouraged policy and before Aborigines-related issues became a popular theme in public discussion. After all, it is very much due to the commitment of Aboriginal elites in the twelve years after the founding (in 1984) of the the first Aborigines' rights group—called the Alliance of Taiwanese Aborigines—that landrights were revised and additionally granted in 1990/91, that the indigenous status of Taiwan's Aborigines was (finally) officially recognized in 1994, and that an "Aborigines representing institution" was established in the central government in 1996 (next to the already existing institution for Mongolians and Tibetans).

2. The officially used term for Taiwanese Aborigines in Taiwan today is *Yuanzhumin* (Autochthones; the abbrevation I will use in the following is YZM). With regard to Xie Shizhong's considerations, I will refer to YZM only in the sense of "Aborigines with wakened ethno-political consciousness," otherwise I will use "Aborigines" as a neutral term (see Xie (1994)). Already in 1984, members of the Alliance of Taiwanese Aborigines (ATA) chose the ethnonym *Yuanzhumin*. Only on July 28, 1994, however, as a result of the third constitutional amendment, was this term officially recognized as substitute for the formerly used term *shanbao* (mountain compatriots). Anthropologists divide the Aborigines living in Taiwan today into at least ten different ethnic groups, all of whose languages belong to the Austronesian language family.
3. For one of the definitions of elites in Taiwan see Chen Ruiyun (1990). Chen refers to all those people as members of elites who have a certain influence on social, political and economic processes. Further, Chen distinguishes (following Pareto (1935)) governing elites and non-governing elites. While he defines those YZM-representatives at the provincial level and above it as governing elites, he refers to well-known members of the YZM-associations and young university and college-intellectuals and ministers as non-governing elites.
4. See http://hledu.nhltc.edu.tw/~tayal/. For Tian, the ethnonym *Tayal* includes members of all three sub-ethnic groups of the Atayal, which are Atayal, Tseole and Sedeq.
5. With respect to linguistic differences, Japanese anthropologists distinguish the three sub-ethnic groups Atayal, Tseole and Sedeq. The latter are again divided into Taroko, Tooda and Dekedaya. Mona Ludao—a chief of the Maxiapo-tribe, which belonged to the Dekedaya, reportedly was the initiator of the Wushe incident in 1930. In that year, Mona was able to mobilize

six of twelve Dekedaya tribes to take part in a collective headhunting raid against the Japanese. This was an act of revenge against the humiliating and dishonoring conduct shown by some Japanese who acted out of ignorance. Though the incident was immediately heavily sanctionized by the Japanese, they were only able to come to terms with Mona Ludao by instigating headhunters of the Tooda and Taroko to attack the Dekedaya (there exists a famous picture of a group of Japanese generals in front of a huge mountain of Atayal skulls). In the following decades, the dislocation/resettlement of the Dekadaya was the only way to keep the three groups from eliminating each other. Since the mid-1980s, the political opposition and members of the Aboriginal movement have attempted to point to Mona Ludao as a symbol of Taiwanese resistance and even of Aborigine resistance against foreign invaders, but only with limited success.

6. Some of the articles provided on the homepage mention the documentary film series "In Search of the Miracles of Taiwan's Tribes" (*Taiwan buluo xunqi*) in Taiwan's cable TV in July 1997. Furthermore, some of the articles selected for the homepage had already been published in Taiwanese newspapers, including one interview with Gimi on the "glory and the responsibility of the 'tattooing culture' " (see " 'Jingmian wenhua' de rongyao yu zeren", in: China Times, July 18, 1997). Another article, originally published in the China Times (on June 20, 1997) and entitled "Taiwan, the Origin of 300 Mio. Austronesians," discusses the visit of Australian archaeologist Peter Bellwood to Gimi's tattooing culture atelier near Hualian.
7. The impression that members of the YZM movement are only poorly supported by the larger Aboriginal society has been certified once again by the election results for YZM legislators in December 1995. Nevertheless, because the DPP allocated one of their seats to non-district representatives (*bufenqu daibiao*) to be occupied by a nominated Aboriginal legislator, the movement could place at least one of their represenatatives.
8. Since the term *bentuhua* is used by both supporters and enemies of Taiwan independence, the scope of this term's interpretative possibilities spans from its being regarded as an "endeavour of taiwanization by a nativist movement of the Taiwanese" to an "endeavour of regionalization within the Chinese province of Taiwan." The author of the comic mentioned here is the Taiwanese caricaturist Qiu Ruolong (1990).
9. Although the office for population registration of the ministry of Interior had "mobilized all forces to encourage Aborigines to rehabilitate their original names" (for example by advertising on all state TV channels or by permitting Aborigines to go through formalities without official identity papers), only 154 of 380 Taiwanese Aborigines had made use of the possibility of name rehabilitation until July 1997. As Lin Jiangyi, director of the YZM-educational board emphasizes, even younger Aborigines (including his own children) do not dare to reveal their YZM status before their social ties and relations in their living environment are fixed (see Taiwan Aktuell Nr, 201, July 16, 1997).
10. It was regarded as suspicious that many members of the Aboriginal elites who claimed that their "mother language was the root of a particular culture

and thus should be practiced more often in everyday life" sent their own children to Han schools in the plains.

11. The intellectuals justified their behavior by emphasizing the importance of opposing these structures, which could take on excessive forms during periods of authoritarian rule.
12. See Keesing (1989).
13. See Keesing (1989: 37).
14. In 1983, shortly after the beginning of the movement for "Taiwanese consciousness," the identity movement of the YZM began to develop. In 1988, the Hakka movement arose; in the early 1990s—already influenced by the claims for multiculturalism—the movement of the so-called "Pingpu" (i.e., those Aborigines who were thought to be assimilated since the early years of the twentieth century) came into existence. In 1993, the plains-dwellers rights movement developed as a strong countermovement of the Han against the YZM. This movement was initiated by those Han who settled in the Aborigines' mountain reservation zones but who did not enjoy any land privileges. In contrast to the land conditions of these Han, YZM land zones were even enlarged in the beginning of the 1990s as a result of the "Return our land movement." Different currents exist within the Aboriginal movement—consider for example the social movement, the literature movement or the cultural movement. I have described these currents in an earlier article (see Rudolph (1996)).
15. Today, the formerly revered slogan *yanhuang zizun* is often parodied, as for instance in the slogan "children and grandchildren of the hundred-paces-snake" (*baibushe de zisun*). Despite the well-known symbolic origin of the hundred-paces-snake (Agkistrodon acutus) as the totem animal of the Paiwan, this slogan was even used by the YZM legislator Cai Zhonghan (who himself is Amis) in 1995.
16. For an understanding of how the "question of provincial origin" (*shengji wenti*) developed into an "ethnic question" (*zuqun wenti*) and finally into a "question of the four great ethnic groups" (*sida zuqun wenti*) see Zhang Maogui (1996a); also see Chang Mao-kuei (1996b).
17. In the literature I have viewed, the terms multicultural and multiculturalist are hardly distinguished; only in rare occasions is *duoyuan wenhua* complimented by *zhuyi* when academics and politicians in Taiwan mean multiculturalism. In most cases (for instance in *duoyuan wenhua zhengzhi and duoyuan wenhu jiaoyu*), the meaning is left to be discerned by the context, and often the English term is added in brackets.
18. See Frank-Olaf Radtke (1993: 81). Here the author contends that a society which considers itself "multicultural" becomes one in which cultural differences of origin take on increasing social and political significance.
19. See Gao Deyi (1995). This discussion touches on the problem of drawing limits between individual and collective rights in a democratic constitutional state.
20. In order to enhance the integration of the people in a multi-ethnic society, the DPP 1993/94 created the term, "fate-community Taiwan," that is "People thrown together by fate" (*Taiwan mingyun gongtongti*), while KMT

politicians rather made use of a more neutral term, that is, *"life-community Taiwan" (Taiwan shengming gongtongti)*.

21. While disregarding the common right to equal opportunity, culturally different populations should be guaranteed an environment in which they have the possibility to increase members of their group. But such a policy also bears the risk of discrimination, insofar as demarcating limits (who should on the grounds of what kind of criteria enjoy what kind of special rights and who should not, or who should be empowered with what kind of authorities and who should not) is always a problem, especially in Taiwan, where ethnicity is generally defined according to the rules of patrilinearity.
22. See Gao Deyi (1995: 14). At least until the mid-1980s, politicians as well as academic elites were strongly convinced that the development of Taiwanese society would proceed according to the models put forward by modernization theory, that is, that social progress would be *a priori* accompanied by a diminution of ethnic, religious and cultural factors. According to this view, the progressive transformation of traditional societal structures and the expansion of "modern" structural elements were an inevitable consequence of functional differentiation and the division of work.
23. In spite of the governments' efforts to quell discontent, officials could not prevent Taiwanese society from voicing intense criticism regarding the education policy, which was widely considered to be both ideology-ridden and low in quality. On April 10, 1994, mass demonstrations for thorough educational reform took place in Taipei. As early as February 1994, people's rights initiatives in Taipei founded the Senlin School (Forest-school). One of its basic maxims was to provide a multicultural education not dominated by one single ideology (see Walisi Yougan (1996)).
24. See Wenjianhui (1995). Lacking other symbols for identification—such as a Taiwanese nation, a homogenous China, or an anti-communist ideology—the reorganization of Taiwanese society along regional and cultural identificatory symbols appeared to be an adequate method to prevent society from further desintegrating. Thus, it does not surprise that the initiative against an "orientation- and identity-vacuum" coincides exactly with the period of time in which Taiwan first relinquished its claim to retake the mainland (in 1992) as well as its claim to be the sole representative of China (in 1994, i.e., by emphasizing that countries that had established diplomatic ties to China were welcome to establish diplomatic ties with Taiwan as well). Besides the endeavor of "community reconstruction," considerations to accelerate the integration of Aborigines into the tourism sector also arose (see Wu Yaofeng (1994)).
25. See *Zhonghua minguo Taiwan Yuanzhuminzu wenhua fazhan xiehui (ZMTYWFX)* 1994: 55 ff.
26. Efforts to implement multicultural education can be first discerned in 1990/91. At that time, the DPP began to introduce lessons for the different vernacular languages of Taiwan in schools of all counties ruled by the opposition party.
27. For differentiation between oppositional or protesting elites (*kangzheng jingying*) and KMT-loyal, political elites (*zhengzhi jingying*) see Xie

Shizhong 1992b. Here, Xie also points to the fact that these categories are gradually disappearing. In her master thesis "Advent of Another World," which was completed in 1997, Wei Yijun (1997) distinguishes between intellectual elites and political elites.

28. See Walisi Yougan (1994).
29. See Gao Deyi (1994).
30. See Wu Tiantai (1993).
31. See Chen Guangxing (1994). Here, Chen also refers to the instrumentalization of the YZM issue in the course of the endeavors of the economically powerful Taiwan to fasten its economic ties with the Malayo-Polynesian peoples of the South Pacific.
32. The cooperation could be observed mainly between members of the cultural movement (artists, writers etc.) and the political elites (local politicians and some of the YZM-representatives in the national representative bodies—Legislative Yuan and National Assembly). The members of the social movement, that is, members of the workers movement, as well as PCT activists, were more reluctant to engage in this kind of realignment, arguing that the Aborigines' social conditions had not changed significantly during the last years.
33. As Stainton (1995) shows in his study on the role of the PCT during the development of the Aboriginal "Return our land" movement, these intellectuals were able to attach themselves to alternative identification-pillars in order to build up a positive self-consciousness: next to ideological patterns inherent in liberation theology, such as the promised land and the chosen people, missionaries and ethnologists also introduced the Fourth World discourse. Stainton worked as a missionary in the Aboriginal section of the PCT from 1980 to 1991.
34. The two festivals were combined into one in order to diminish the frequency of home travel for the participants, who mostly worked in the big cities far away from Qimei. Another reason for combining the two festivals was, of course, the desire to cater to the seasonal needs of tourism.
35. See Xie Shizhong (1992a); also see Hsieh Shih-chung (1994).
36. For a more thorough discussion see Rudolph (1993, 1994).
37. During the 1980s, the legend of Wu Feng was the most well-known symbol of the civil inferiority and backwardness of Taiwanese Aborigines. The legend tells of the honorable death of a Han Chinese who, loyal to his Confucian principles, had dedicated himself to the task of moral improvement among the Aborigines. To protect his Han compatriots, he finally sacrificed himself as prey to the headhunters. Studies of Taiwanese ethnologists show that although the legend already existed in Taiwan during the Qing (1683–1895), it was first propagated extensively in the period of Japanese colonial rule, when Wu Feng was reinterpreted as a symbol of virtuous self-sacrifice (Wu Feng as "Christ of the East"). After 1945, Jiang Kai-shek ordered that the legend be part of the curriculum in elementary schools, and a memorial statue was built in Jiayi. The inscription of the statue points to the noble spirit of Wu Feng, who was said to be a martyr and the savior of the Han (see Guan Hongzhi (1987)).

38. In an article on multicultural education in Taiwan's elementary schools, ethnologist Wu Tiantai asserts that fights and disagreements between Taiwan's different peoples (e.g., the debates surrounding the Wu Feng story) had best not be mentioned any longer in the curriculae for multicultural education to avoid the generation of negative stereotypes (see Wu Tiantai (1993)).
39. *Yuanzhumin wenhua chucao xuanyan* in Chinese. With the exception of the Yami, all of Taiwan's Aboriginal groups practiced headhunting (*chucao* in Chinese) until the beginning of the twentieth century.
40. The first official recognition of the ethnonym *Yuanzhumin* occurred during the speech Li Denghui gave on the last day of the congress. For a more detailed descripti on of the event see Rudolph (1996).
41. See Walisi Yougan (1992). Because of the similarity of the Chinese characters, the title of the book (*Fandao chuqiao*) easily evokes the association of "Headhunting with the Savages' Knife" (*fandao chucao*).
42. Several YZM-authors referred to the cultural relativism of the historian Dai Guojun who argued that the Han also killed human beings in war or skirmishes and that it was thus unfair to label Taiwanese Aborigines as raw-natured and wild just because they reacted with "*chucao*" when threatenend by the Han. Hence, Wadan argues that here we merely see "variations in the style of killing" (see Watan Baser (1987)).
43. See Fanon (1961).
44. See Qiu Ruolong (1990). Because of high demand for this book, it was published once again in 1995 at the sixty-fifth memorial of the Wushe incident (1926.10.).
45. See Qiu Ruolong (1994).
46. Already, in 1991, Qiu's comic illustrations of the tattooing-culture were published in a volume of Atayal-myths written by two Atayal intellectuals in the Chinese/Atayal language. The popularity of Qiu in the circles of the Atayal-Elite was attested again in 1996 with the inclusion of his illustrations in the book by Liyiging Yuma entitled "Tradition" (*Chuancheng*), where the young Atayal woman writer discusses the hardships, frustrations and tokenism of the *Yuanzhumin* movement.
47. See China Times (October 27, 1995). The articles are accompanied by copies of headhunting photographs from the Japanese colonial period. These photographs were borrowed from the collection of materials assembled for the commemorative ceremony.
48. See Chen Shaoru (1994).
49. See Jiang Guanming (1994).
50. See Chen Zhaoying (1995).
51. See *Taiwan shiliao yanjiu*, No. 2, Taibei, 1993: 28/26. This same perspective became manifest in the symposium held by Chen Fangming, Zhang Yanxian, Zheng Liangwei and other important personalities of the Taiwanization-movement.
52. See Sun Dachuan (1993).
53. See Fu Dawei (1993).
54. "*Bailang*" is the Mandarin transcription for the Atayal-pronounciation of "bad person" in the Hokkien dialect.

55. In the article entitled "Taiwan's Ethnic Politics in Transformation," DPP-member Yang Changzhen contends: "From the perspective of the protagonists of the *Yuanzhumin* movement, the different Aboriginal groups attempt to present an image of a [unified] group of Taiwan Aborigines. Yet from the perspective of the people in the mountains, in the countryside and in the tribes, for instance, for the old grandfather or the old grandmother, a Taiwan-aborigine ethnic group, does not exist: this is an empty term. In the actual society of Taiwan, however, having an ethnic identity is a must—only through the formation of an ethnic group, can one attain status in political interactions" (see Yang Changzhen (1995)).
56. In the same article as mentioned earlier, Yang emphasizes how necessary it was for the political opposition in Taiwan to set up a moral codex and a value system that was independent of China and its traditional cultural maxims: by refering to "orthodoxy," the KMT could, with one sweep, not only legitimize its prerogative to rule, but also its demand for utmost loyality from its citizens. Everything that threatened "orthodoxy" was pejoratively labeled "rebellion." Caught within this value frame set up by the KMT in Taiwan, the political opposition first tried to be even more "orthodox" than the government in order to overthrow it, yet it failed because rebellion against "orthodoxy" was to them an unacceptable concept. At last the opposition understood that as long as they argued on the same level of "orthodoxy," no significant political breakthrough could be achieved. For this reason, the opposition turned to a new Taiwanese orthodoxy which could legitimize a different moral and ethical system (see Yang Changzhen (1995: 203)).
57. See the reprint of Xie Shizhong's comments on the nature of ethnic symbols in one of the pamphlets of the ATA (including symbols as pan-ethnic name, ethnic names, individual Aboriginal names, mythical heros, historic events, art and cloth (see ATA/PCT (1992)).
58. See Liu Shaohua (1993).
59. See Werner Schiffauer (1997).
60. One of the examples refered to by Radtke (1993) is the dispute about the wearing of veils by Islamic women in Europe, a dispute in which the national dictum of secularism stands in non-unifiable confrontation with religious fundamentalism.
61. That is, local people's representatives (which may also be Han) and failed candidates of local political elites, who have exert to extraordinary measures to increase their popularity.
62. With many of these points, it is probably possible to find compromises in the course of time; for instance, in one of his articles, Taiwanese anthropologist Xie Shizhong tries to persuade ethnic elites that the hybridization of ethno-cultural symbols might also be helpful for the development of a pan-ethnic Yuanzhumin movement (see Xie (1992a)).
63. See Radtke (1993). There is also a danger that conflicts regarded as ethnic in nature will be compensated regressively.
64. The point of departure for these reflections is the problem of demarcating boundaries between individual and collective rights in a multiculturally oriented democratic state. As Taylor makes clear, the rights of the individual

generally stand in the foreground in democratically conceived societies. These rights are guarded by a neutral state that has no cultural or religious ambitions of its own nor collective goals beyond the sustenance of the personal freedom and physical protection of its citizens—that is, no such aims beyond the well-being and safety of its members. The recognition of common human dignity stands as the highest principle of this universalist approach. The adherents of the culturalist approach argue that in an era during which erstwhile reciprocity based social bonds have waned and in which the satisfaction of individuals' basic need for recognition is hence no longer a foregone conclusion, a policy merely geared toward the recognition of universal human dignity cannot suffice. Equally important is the recognition of the unique, unmistakable identity of any individual; in short, the recognition of difference and of variable environmental contexts (i.e., the collective from which an individual came and to which he/she binds his/her identity (see Taylor (1992)).

Bibliography

ATA (Alliance of Taiwanese Aborigines), 1987, *Yuanzhumin—bei yapozhe de nahan [Taiwanese Aborigines—The Cry of the Oppressed]*, Taiwan Yuanzhuminzu quanli cujinhui chengli sanzhounian zhuanji, Taibei 1987.

ATA/PCT (Presbyterian Church of Taiwan), 1992, *Yuanzhumin xuandao weiyuanhui, 1992, Zhengqu xianfa "Yuanzhuminzu tiaokuan" xingdong shouce (Booklet on the Strive for a "Paragraph on Taiwanese Aborigines" in the Constitution)*, (PCT) Taibei 4/1992: 13.

Chang Mao-kuei, 1996b, "Political Transformation and the 'Ethnization' of Politics in Taiwan" in: Schneider, Axel and Guenter Schubert (eds.), *Taiwan an der Schwelle zum 21. Jh.—Gesellschaftlicher Wandel, Probleme und Perspektiven eines asiatischen Schwellenlandes*, Mitteilungen des Instituts fuer Asienkunde Hamburg vol. 270, Hamburg 1996: 135–152.

Chen Guangxing, 1994, "Diguo zhi yan: 'ci' diguo yu guozu—guojia de wenhua xiangxiang" (The Imperialist Eye: The Cultural Imaginary of a Sub-Empire and a Nation State), in: *Taiwan shehui yanjiu jikan*, No. 17, Taibei 7/1994: 149–222.

Chen Ruiyun, 1990, "Zuqun guanxi, zuqun rentong yu Taiwan Yuanzhumin jiben zhengce" (Ethnic Relations, Ethnic Identity and Aboriginal Policy in Taiwan), non-published M.A thesis, Zhengzhi University, 1990: 29–33.

Chen Shaoru, 1994, "Shilun Taiwan renleixue de Gaoshanzu yanjiu" (Preliminary Discussion of the Gaoshanzu Research in Taiwan's Cultural Anthropology), in: *Shanhai wenhua zazhi*, No. 6, Taibei 11/1994: 27–36.

Chen Zhaoying, 1995, "Lun Taiwan de bentuhua yundong: yi ge wenhuashi de kaocha" (Discussion of the Taiwanization Movement: Examination of a Cultural History), in *Zhongwai wenxue*, Vol. 23, No. 9, 2/1995: 8–43.

Fanon, Frantz, 1961, *Les damnées de la terre*, Paris 1961.

Fu Dawei, 1993, "Bailang senlin li de wenzi lieren" (Hunters of Chinese Characters in the Forest of Han Rascals), in *Dangdai zazhi*, No. 83, 3/1993: 28–49.

Gao Deyi, 1993, "Maixiang 'duoyuan yiti' de zuqun guanxi: Yuanzhumin jiben zhengce de huigu yu zhanwang" (Towards a "pluralist entity" in Ethnic Relations: Review and Prospects of Aboriginal Policy in Taiwan), in *Zhonghua minguo Taiwan Yuanzhuminzu wenhua fazhan xiehui (ZMTY-WFX), Yuanzhumin zhengce yu shehui fazhan*, Taibei, 1994: 140–188.

Gao Deyi, 1995, "Maixiang duoyuanhua jiaoyu: Yuanzhumin jiaoyu xiangguan fagui de jiantao" (Towards a Pluralist Education: Critical Review of the Regulations Aboriginal Education in Taiwan), in *Yuanzhumin jiaoyu yantaohui*, Hualian shifan xueyuan 1995: 12–32.

Guan Hongzhi, 1987, "Minzhong de Wu Feng lun" (The Public Discourse on Wu Feng), in *Renjian* 8/1987.

Hsieh Shih-chung, 1994, "Tourism, Formulation of Culture, and Ethnicity: A Study of the Daiyan Identity of the Wulai Atayal", in Harrell, Stevan and Huang Jun-chieh (eds.), *Cultural Change in Postwar Taiwan* (Westview Press), Colorado 1994: 184–201.

Jiang Guanming, 1994, "Chucao xuanyan shi Yuan/Han duihua de qidian—ping 1994 Yuanzhumin wenhua huiyi" (The Headhunting Raid-Manifest is the Starting Point of a Dialogue between Taiwanese Aborigines and Han: A Comment on the Aboriginal Culture Conference in Taiwan 1994), in: *Shanhai wenhua shuangyuekan*, No. 6, Taibei 9/1994: 37–44.

Keesing, Roger M., 1989, "Creating the Past. Custom and Identity in the Contemporary Pacific", in *The Contempory Pacific* I (1 and 2) 1989: 19–42.

Li Yiyuan et al., 1983, *Shandi xingzheng zhengce zhi yanjiu yu pinggu baogaoshu (Evaluation Report on Taiwan's Aboriginal Policy)*, Academia Sinica, Taibei 1983.

Liu Shaohua, 1993, "Yuanzhumin wenhua yundong de lishi weizhi" (The Historical Position of the Culture Movement of Taiwanese Aborigines), in: *Shanhai wenhua shuangyuekan*, No. 1, Taibei 11/1993: 48–55.

Qiu Ruolong, 1990, *Wushe shijian [The Wushe Incident]*, Shibao wenhua chuban, Taibei 1990 (2nd ed. 1995).

Qiu Ruolong, 1994, "Tayazu de jingmian wenhua" (The Tattooing Culture of the Atayal), in: Tayazu renwen lishi yantaohui: Taya wenhuaji 5/1994 (quanguo wenyiji), Taiwan 1994: 12.

Radtke, Frank-Olaf, 1993, "Politischer und kultureller Pluralismus. Zur politischen Soziologie der 'multikulturellen Gesellschaft' " (Political and Cultural Pluralism: On the Political Sociology in a "Multicultural Society"), in: Robertson-Wensauer, Caroline Y., *Multikulturalitaet—Interkulturalitaet? Probleme und Perspektiven der multikulturellen Gesellschaft*, (Nomos-Verlag) Baden-Baden 1993: 89.

Rudolph, Michael, 1993, *Die Prostitution der Frauen der Taiwanesischen Bergminderheiten—historische, sozio-kulturelle und kultur-psychologische Hintergruende (Taiwan's Aborigines and the Prostitution Problem—Historical, Socio-Cultural and Psycho-Cultural Backgrounds)*, (LIT Verlag) Hamburg/Muenster 1993.

Rudolph, Michael (Liu Zhexun), 1994, "Taiwan shehui bianqian de shaoshu minzu funue changji wenti—shehui wenhua, shehui xinli, ji lishixing de yinsu" (Social Change in Taiwan and the Prostitution Problem of Taiwan's

Aboriginal Women—Socio-Cultural, Psycho-Cultural and Historical Factors), in *Taiwan Indigenous Voice Bimonthly*, Nr. 4, Taipei 1994.

Rudolph, Michael, 1996, "'Was heisst hier 'Taiwanesisch'—Taiwans Ureinwohner zwischen Diskriminierung und Selbstorganisation" (Who has the Right to call himself 'Taiwanese'?—Taiwan's Aborigines between Discrimination and Self-Organization), in Schneider, Axel u. Gunter Schubert (ed.), *Taiwan an der Schwelle zum 21. Jh.—Gesellschaftlicher Wandel, Probleme und Perspektiven eines asiatischen Schwellenlandes*, Mitteilungen des Instituts fuer Asienkunde Hamburg vol. 270, Hamburg 1996: 285–308.

Stainton, Michael, 1995, *Return our Land: Counterhegemonic Presbyterian Aboriginality in Taiwan*, York University 1995.

Sun Dachuan, 1993, "Yuanzhumin wenxue de kunjing—huanghun huo liming" (The Dilemma of Aboriginal Literature in Taiwan—Dusk or Dawn), in: *Shanhai wenhua shuangyuekan*, No. 1, Taibei 11/1993: 97–105.

Taylor, Charles, 1992, *Multiculturalism and "The Politics of Recognition,"* Princeton 1992.

Walisi Yougan, 1992, *Fandao chuqiao (Drawing the Savages' Knife)*, Daoxiang cbs, Taibei 12/1992.

Walisi Yougan, 1994, "Yuyan, zuqun yu weilai: Taiwan Yuanzhuminzu muyu jiaoyu de ji dian sikao" (Language, Ethnic Groups and Future Prospects: Some Reflections on Vernacular Language Education of Taiwan's Aborigines), in: *ZMTYWFX*, Yuanzhumin zhengce yu shehui fazhan, Taibei 1994:190–221.

Walisi Yougan, 1996, "Cong xianxing de jiaoyu zhengce kan muyu jiaoyu—yi Yuanzhumin jiaoyu zhengce wei zhu" (Looking at the Vernacular Language Education Policy of Taiwan's Aborigines), in: *Jiaoshou luntan zhuankan*, Taibei 1996: 417–431.

Watan Baser, 1987, "'Chucao huo fanhai' shi Yuanzhumin kang Han de yiju" ["Headhunting Raids" were the Method of Resistance of Taiwanese Aborigines against the Han], in: *ATA (Alliance of Taiwanese Aborigines), 1987*, Yuanzhumin—bei yapozhe de nahan, Taiwan Yuanzhuminzu quanli cujinhui chengli sanzhounian zhuanji, Taibei 1987: 153–160.

Wei Yijun, 1997, "Ling yi ge shijie de lailin: Yuanzhumin yundong de lilun shijian" (Advent of Another World: Realization of Theory in the Aboriginal Movement in Taiwan), M.A. dissertation, National Tsinghua University, Taiwan 1997.

Wenjianhui 1995.

Werner Schiffauer, 1997, "Kulturdynamik und Selbstinszenierung—Kulturalismus im postnationalen Zeitalter: Sich als Gruppe konstituieren und Gehoer verschaffen" (Dynamics of Culture and Self-Inscenation—Culturalism in the Post-National Era), in: *taz* 4.3.1997: 14f.

Wu Tiantai, 1993, "Xiaoxue de duoyuan wenhua jiaoyu" (Multicultural Education in Taiwan's Elementary Schools), in: Zhongguo jiaoyu xuehui, Taiwan shudian 1993: 375–386; also as: Wu Tiantai, 1996, "Xiaoxue duoyuanhua jiaoxue", in: Cai Zhonghan, 1996, *Yuanzhumin xiandai shehui shiying* 2, Jiaoyu guangbo diantai, Taibei, 1996: 626–657.

Wu Yaofeng, 1994, "Yuanzhumin wenhua ziyuan zhi fajue" (The Unearthing of Aboriginal Culture Ressources in Taiwan), in: *83 niandu shandi yanxiban jiangyi*, Taibei 6/1994.

Xie Shizhong, 1992a, "Guanguang huodong, wenhua chuantong de sumo, yu zuqun yishi: Wulai Daiyazu Daiyan rentong de yanjiu" (Tourism, Formulation of Culture, and Ethnicity: A Study of the Daiyan Identity of the Wulai Atayal), in *Kaogu renleixue kan*, No. 48, Taibei 12/1992: 113–129.

Xie Shizhong, 1992b, "Pianli qunzhong de jingying: Shilun 'Yuanzhumin' xiangzheng yu Yuanzhumin jingying xianxiang de guanxi" (Elites without People: A Preliminary Discussion of the Elites Phenomenon of Taiwanese Aborigines), in: *Daoyu bianyuan*, No. 5, Taibei 10/1992: 52–60.

Xie Shizhong, 1994, "Shandi gewu zai nar shangyan" (Where shall the Songs and Dances of Taiwanese Aborigines be performed), in: *Zili zaobao* 19.12.1994.

Xu Muzhu et al., 1992, *Shanbao fudao cuoshi jixiao zhi yantao (Discussion of the Effects of the Assisting Measures for Taiwan's Aborigines)*, Academia Sinica, Taibei 1992.

Yang Changzhen, 1995, "Zhuanxingqi de Taiwan zuqun zhengzhi" (Taiwan's Ethnic Policy in Transition), in: PCT Taizhong zhonghui, 1995, *Taiwan xinian-qiguo xilie yantao zhuanji*, Taizhong 1995: 187–220.

Zhang Maogui, 1996a, "Taiwan zui cishou de zhengzhi wenti" (The Most Critical Political Problem in Taiwan), in: *Cai Xun (Wealth Magazine)* No. 168, 3/1996: 152–163.

Zhonghua minguo Taiwan Yuanzhuminzu wenhua fazhan xiehui (ZMTY-WFX), *Yuanzhumin zhengce yu shehui fazhan (Taiwan's Aboriginal Policy and the Development of Aboriginal Society in Taiwan)*, Taibei 1994.

CHAPTER 6

IDENTITY POLITICS AND THE STUDY OF POPULAR RELIGION IN POSTWAR TAIWAN

Paul R. Katz

One of the most striking developments in the study of popular religion on the island of Taiwan has been the increasing emphasis on the ways in which local beliefs and practices may or may not contribute to the formation of a Taiwanese identity. Deep-seated resistance to Nationalist Party (KMT) education and cultural policies, as well as a growing interest in Taiwan's own history and culture have prompted a growing number of Taiwanese scholars to study popular religion.[1] At the same time, however, research on Taiwanese popular religion has become enmeshed in an increasingly lively (and sometimes bitter) debate concerning the "Taiwaneseness" (or "Chineseness") of the island's culture. Like other debates about identity the world over, this discourse has centered less on discernible and objective distinctions between two groups of people (in this case mainland Chinese and Taiwanese) but on perceived differences often inextricable from the realm of sentiments and beliefs (Anderson 1991; Holcombe 1995). One interesting example of the impact of identity politics on Taiwan's scholarly community has been the intense debate in 1997 about the contents of a new series of textbooks to be used in Taiwan's middle schools entitled *Getting to Know Taiwan* (*Renshi Taiwan*), especially the volumes on history and social studies. On the one hand, scholars favoring a pro-China or reunification agenda have criticized these books for advocating Taiwanese independence, painting too rosy a picture of the Japanese Occupation era (1895–1945), and ignoring Taiwan's cultural links to China. On the other hand, scholars advocating a pro-independence stance or at the very least opposed to reunification have claimed that these books do not go far enough in fostering a sense of Taiwanese identity.[2]

The debate over identity has also affected the study of a highly important problem: the extent to which the beliefs and practices brought to Taiwan by migrants from mainland China may have experienced a process of "indigenization" (*bentu hua; tuzhu hua*) or "Taiwanization" (*Taiwan hua*). Some papers presented at academic conferences over the past few years have been noteworthy in their attempts to define Taiwanese popular religion as a cultural phenomenon unique to Taiwan. These papers also view popular religion in Taiwan as being based on a sense of identity that excludes mainland China as a source of cultural tradition. Whether this new sense of identity has gained widespread acceptance among the people of Taiwan, or has only been embraced by some of the island's intellectuals and politicians, remains to be determined. Nevertheless, the impact of such arguments on academic discourse has been considerable, with an increasing number of scholars attempting to find ways in which Chinese popular religion (particularly in China's southeastern coastal regions) adapted to Taiwan's unique historical conditions.[3] All this reveals the considerable impact identity politics have had on Taiwan's academic discourse, and suggests that the study of both the politics and relevant scholarship will be of prime importance in the near future.

This essay attempts to explore current academic discourse concerning the "indigenization" of popular religion in Taiwan. It does so by means of a case study of the cult of the Royal Lords (*wangye*). The focus of the chapter will be on this cult's history in Fujian and Taiwan during the Qing dynasty, although data from the Japanese occupation and postwar eras will also be discussed. In Taiwan and parts of southern China, worshippers have used the title "Royal Lord(s)" to refer to a wide range of spirits, including plague-spreading deities belonging to the celestial bureaucracy's Ministry of Epidemics (*Wenbu*), and vicious ghosts (*ligui*) of various unruly spirits awarded the title "king" (*wang*) because of their ability to control the ravages of contagious diseases. Such cults appear to have developed in south China by the tenth and eleventh centuries, and gained increasing popularity after the twelfth century. The most widely worshipped Royal Lord in south China and Taiwan is a deity known as Marshal Wen (*Wen Yuanshuai*), who is worshipped in southern Fujian and Taiwan as Lord Chi (*Chi Wangye*). One of the most striking aspects of the Royal Lords cults involves the performance of plague expulsion festivals, either as a prophylactic measure or when an epidemic was in progress. Such festivals usually conclude with the expulsion of a boat known as a "plague boat" (*wenchuan*) or "Royal Lords boat" (*wangchuan*), which represents the community's accumulated afflictions. Different types of boat expulsion rituals have been held in south China and Taiwan for centuries, and are still popular today.

Immigrants from Fujian brought the Royal Lords cult to Taiwan during the seventeenth and eighteenth centuries. According to Liu Chih-wan, a total of 74 Royal Lords temples large enough to be recorded in local gazetteers were founded in Taiwan during the Qing dynasty. Cults to the Royal Lords continued to grow in Taiwan during the Japanese occupation period, with a 1918 temple survey by the colonial authorities listing a total of 447 registered temples to these deities.[4] The Japanese colonial authorities largely ignored cults to the Royal Lords and other Taiwanese local deities until the Kominka (Japanization) Movement (1937–1945), when some temples and/or the statues therein were destroyed. However, the effects of this repression do not appear to have been too severe, and the number of registered Royal Lords temples actually increased from 534 to 677 between the 1930 and 1960 surveys.

Today, Lord Chi remains the most popular Royal Lord worshipped in Taiwan. Of Taiwan's nearly 800 registered Royal Lords temples, at least 131 (including 43 in Tainan County) enshrine Lord Chi as their main deity (*zhushen*), while hundreds more include him as one of their subsidiary deities (*peishen*). Another well-known deity is Lord Wen (Wen Wangye), whose cult appears to be related to that of Marshal Wen. He is the main deity of over 60 Royal Lords temples in Taiwan, the largest and most popular of which is the Palace of Eastern Prosperity (Donglong Gong), located in East Haven (Donggang; a fishing port in Pingdong County). Lord Wen's cult has continued to spread throughout the island during the past few years.[5]

The chapter begins with an analysis of the factors contributing to identity formation in Taiwan, as well as the complex role popular religion has played in this process. I then explore current academic discourse in Taiwan that attempts to show that the history of the Royal Lords cult has been shaped by processes of indigenization. I wish to emphasize that the goal of this chapter is *not* to determine whether or not the cult of the Royal Lords actually "indigenized" after it took root in Taiwan. As I shall argue below, the data gathered to date is insufficient to adequately address this complex issue. I simply attempt to determine the impact of identity politics on postwar scholarship about Taiwanese popular religion by critically examining the arguments made in favor of "indigenization" of the Royal Lords cult, as well as the sources they are based on.

Identity Formation in Postwar Taiwan

At present, the most systematic studies of identity formation in Taiwan have been undertaken by Thomas B. Gold and Alan M. Wachman (Gold 1994; Wachman 1994a,b), two scholars who view the process of identity formation in Taiwan from very different perspectives. Gold argues that "questions of identity were raised as part of the strengthening and politicization of civil society" (Gold 1994: 59), and adopts Antonio Gramsci's definition of civil society for capitalist societies (which Taiwan is not), where it constitutes a "superstructural realm of social relations and organizations" through which the elite attempt to exert their dominance over various facets of social life (51). In the case of Taiwan, Gold describes the rise of a new bourgeois elite composed mainly of Taiwanese beginning mainly in the 1980s, and demonstrates that members of this elite have often challenged KMT hegemony over cultural affairs, including the production of national and/or ethnic identity (50–53).

For his part, Wachman also sees identity formation as linked to power struggles between different groups in Taiwan (Wachman 1994b: 26–28, 33, 38, 49, 62). He also claims that KMT policies enacted during the early decades of its rule over Taiwan, which were aimed at denying Taiwanese self-determination, ended up playing a key role in recent formations of local identity (27, 40–42, 48–49, 60, 62). Wachman argues that:

> Essentially, Taiwanese culture is a regional variation of Chinese culture. It is not wholly unique and shares a good deal with the culture of southeastern China, particularly Fukien province, across the Taiwan Strait. Yet, the identity Taiwanese feel and the idea and the reason why some have tried to promote the idea that Taiwan has a separate culture has to do with Taiwanese reactions to political repression. (49)

Wachman carefully analyzes five key factors that he believes have contributed to the development of Taiwanese identity, including: (1) Relative geographic isolation plus a history of being ruled by outsiders; (2) Tension caused by KMT attempts to impose its own version of a national elite culture; (3) Collective memories of confrontations such as the 228 Incident; (4) Mistrust and even hostility towards the KMT; and (5) KMT policies designed to "institutionalize" differences between Taiwanese and Mainlanders (42–52, 62).

In many ways, Gold's and Wachman's analyses effectively represent two sides of the same coin. Gold has convincingly demonstrated that many of the island's upwardly mobile elites have been actively promoting a sense Taiwanese identity. At the same time, however, many of these individuals feel that they were denied such an identity during their youth, particularly while studying at the elementary and middle-school levels (Wachman: 40–41, 55; Lin 1987; Tu 1998: 149–161; Wilson 1970). The resulting sense of betrayal, combined with pride in their Taiwaneseness and concern about its preservation, have led many members of Taiwan's elite to play an active role in current identity politics (Wachman 1994b: 51–56). Wachman's work is particularly important because he explores the concept of identity in general, and the factors that help shape identity in particular. For example, Wachman points out that while identity is largely an "artificial" notion connected with "sentiments and beliefs," it can become a highly potent issue when combined with political objectives (28). He also notes that identity is not a natural state of consciousness, but needs to be created (or invented). However, while those who are promoting visions of identity attempt to achieve a state of collective consciousness, individual notions of identity may differ greatly (28).

One aspect of Taiwanese culture that neither Gold nor Wachman discuss in detail is popular religion. This is a pity; for it is clear that popular religion has also played an important role in the attempt to create a Taiwanese identity. As scholars like Emily Martin [Ahern], Hill Gates [Rohsenow], and Robert Weller have shown, the Taiwanese frequently relied on local cults and festivals to express tacit resistance toward KMT rule as early as the 1950s and 1960s (Ahern 1981; Rohsenow 1973; Weller 1987, 1994). Weller's fascinating case study of the cult of the Eighteen Lords (Shiba wanggong) in postwar Taiwan, describes how the spirits of seventeen men and their loyal canine companions currently receive offerings of cigarettes and other items from people wishing to make a quick profit, including prostitutes and members of Taiwan's criminal underworld. Efforts by the state and even the temple committee to mold popular opinion proved largely fruitless, mainly because temples in general usually prove unable to create "strong social relations of interpretation" to support their views (Weller 1994: 169; see also Weller 1996: 156–157). Weller's analysis is highly relevant for this study, as it shows the key role that popular religion could play in resisting KMT attempts at imposing cultural hegemony. As a result, Joseph Bosco has stressed the importance of popular religion in identity formation, and notes that the KMT's decision to support popular religion rather than suppress it represents an effort to claim it as Chinese and deny its Taiwaneseness (Bosco 1994). However, using popular religion as a means of promoting identity is not without its pitfalls, regardless of whether the identity in question is Chinese or Taiwanese. In particular, the opening of relations between China and Taiwan has allowed Taiwanese to go on pilgrimages to the mainland, thereby embracing a broad sense of Taiwanese cultural identity that encompasses both sides of the Taiwan Straits (Wachman 1994b: 57). While some scholars have seen Taiwan's popular religion as contributing to the formation of an emerging Taiwanese identity that distances itself from mainland China, others such as Joesph Bosco claim that religion and other facets of Taiwan's popular culture serve to transcend such divisions (Bosco 1994). Steven Sangren presents an even more complex picture, arguing that Taiwanese tourists and pilgrims journeying to Fujian may be involved in seeking objectifications of identity, and that for many worshippers there appears to be no contradiction between a popular deity's role as a symbol of Taiwanese identity and its links to a broader definition of identity which includes Fujian, Taiwan, and some overseas Chinese communities throughout Southeast Asia (Sangren 1996: 15–18). Kenneth Dean also notes that access to Fujian's local cults can further increase "the range and flexibility" of Taiwanese beliefs and practices (Dean 1998: 271).

The Royal Lords and Taiwanese Identity

Thus far this chapter has focused on the problem of how research on Taiwan's popular religion (in this case cults to the Royal Lords) has been affected by the recent political and socioeconomic changes taking place in Taiwan. During the past few years, a number of scholars have presented papers at local academic conferences, including two of which will be discussed below, which are noteworthy in their attempts to define Taiwanese popular religion as a cultural phenomenon unique to Taiwan. These scholars also view Taiwanese popular religion as being based on a sense of identity that excludes mainland China as a source of cultural tradition. It is interesting to note that in this respect their arguments differ significantly from those of some American anthropologists working in postwar Taiwan, who noted the importance of popular rituals as a form of resistance to KMT rule but did not attempt to prove that the contents of such rituals were different from those in China (see earlier).

Most research on the problem of "indigenization" of popular religion in Taiwan has centered on the cult of Mazu (see, for example, Rubinstein 1993, 1995; Sangren 1996). However, two recent papers by Tung Fang-yüan and Li Feng-mao have attempted to prove that "indigenization" also occurred in the case of the cult of the Royal Lords. Tung's paper, which deals with Taiwanese popular religion as a whole, represents the work of a scholar motivated by a strong commitment to define where the lines of legitimate cultural (not to mention political) identity need to be drawn. Li also shows an interest in the issue of identity, but also emphasizes the role of Taiwanese popular religion and local culture as a whole in providing a stable foundation for modern Taiwanese society.

Tung clearly links popular religion with the issue of ethnicity, arguing in the first paragraphs of his paper that "Taiwanese folk beliefs represent the traditional religion of Taiwan's Hokklo and Hakka peoples" (Tung 1995: 809), and that "Taiwanese folk beliefs are the traditional religion of the [island's] Fujianese and Cantonese ethnic groups (*Min-Yue zuqun*)" (809–810).[6] He also reveals his agenda at a number of places in the paper, particularly the conclusion, insisting that one of the missions (*shiming*) of popular religion in postwar Taiwan involves the preservation of Taiwanese culture and "Taiwanese consciousness" (*Taiwan yishi*). In addition, Tung also stresses that the process of "indigenization" (*bentu hua*) in cultural arenas such as popular religion can help dissolve the erroneous notion that "blood is thicker than water" (*xue nong yu shui* (828).[7] Tung's comments are in all likelihood motivated by a desire to overturn the arguments of mainland Chinese scholars who support PRC reunification policy by arguing that Fujianese and Taiwanese popular religion are identical and represent a form of cultural unity.[8]

Tung argues that Taiwan's Royal Lords have experienced "indigenization" in the following two ways. First, he claims that from the Qing Dynasty onward the Royal Lords became transformed from roving plague spirits (*wenshen*) to all-powerful deities who settled down on Taiwan and brought blessings to their worshippers (819). Later on, he states that the practice of expelling plague deities by means of sending off a Royal Lords boat changed from floating small boats to burning larger boats, in part because boat-floating rites had been forbidden by local officials (821). The validity of these arguments is discussed in the following paragraphs.

As for Li Feng-mao, his main concern involves using local gazetteers to speculate on how processes of "indigenization" (which he prefers to label *bendi hua*) have shaped the growth of Taiwanese popular religion, including the cult of the Royal Lords. He opens his paper with a review of Li Kuo-ch'i's concept of "inlandization" (*neidi hua*); changes in Taiwan similar to those in mainland China and Ch'en Ch'i-nan's concept of "indigenization" (*tuzhu hua*).[9] Li Feng-mao attempts to use both of these concepts in his analysis of Taiwanese history, arguing that "inlandization" affected politics and education during the early period of Qing rule over Taiwan, while "indigenization" influenced the island's subsequent cultural development (Li 1995: 835). Such an argument makes sense in that it does not treat the concepts of "inlandization" and "indigenization" as being mutually exclusive historical phenomena. However, whether they occurred in succession, and directly impacted the growth of the Royal Lords cult on Taiwan is much less certain, as we shall see later.

Li opens his paper with a one-paragraph overview of Royal Lords cults and rituals in Fujian, based on the results of a partial survey of that province's local gazetteers (Li 1995: 839). He also claims that during the initial growth of the Royal Lords Taiwan was marked by the presence of "inlandization" (840). However, he then proceeds to argue that Taiwan's socioeconomic stability during the eighteenth and nineteenth centuries, combined with the decline of sub-ethnic feuds (*xiedou*),[10] resulted in the "indigenization" of cults like the Royal Lords. Li cites a number of factors he believes reflect the presence of this process, beginning with the materials used to build the Royal Lords boat. According to Li, unspecified "material conditions" in early Taiwan caused artisans to stop building such boats out of wood and instead shift to using paper and/or bamboo. He also indicates that the supposed change from floating Royal Lords boats to burning them may be linked to these material conditions, as wooden boats should float better than ones built of paper or bamboo (842–884). There are two problems with this argument. One is that the exact times at which different communities of Taiwanese people modified local boat expulsion rituals, and in what ways, are extremely difficult to determine. Furthermore, any arguments that material conditions (such as a shortage of wood or trained artisans to carve it) may have influenced ritual practice need to be buttressed by data on Taiwan's economic and/or environmental history.[11]

Li also attempts to show that "indigenization" occurred during the Japanese Occupation period. He pays particular attention to the colonial government's disease prevention efforts, arguing that the success of these policies influenced the cult of the Royal Lords. In particular, he draws on the work of Liu Chih-wan to claim that the practice of floating Royal Lords boats finally ceased for good in large part due to public health regulations forbidding the practice (848). Li is correct in stressing that colonial policies may have affected the growth of the Royal Lords cults in general, and the performance of boat expulsion rites in particular. However, the disease history of that period is only now being studied in detail (Fan 1994; Katz 1996), and little data has come to light that could help determine the ways in which the colonial government's public health policies influenced cultic development. In addition, although a growing body of research has begun to treat the religious policies enacted during the Japanese occupation (see, for example, Miyamoto 1988 and Ts'ai 1994), scholars working on this topic have yet to discover any regulation banning the floating away of Royal Lords boats.

Li maintains that "indigenization" has shaped the ways in which popular representations of the Royal Lords and boat expulsion rituals have changed over time. Based on reports by Japanese ethnologists and officials (see, for example, Kataoka 1981 (1921); Maejime 1938), Li concludes that by the early twentieth century the Royal Lords boat had become an auspicious symbol, as opposed to a portent of impending calamity (Li 1995: 849). He makes similar arguments about the postwar era, claiming that Taoist Plague Offerings (*wenjiao*) performed during that time have been transformed from fearful exorcistic rites into highly festive occasions (857). Li also asserts that popular images of the Royal Lords changed from deities charged with protecting seafarers to those especially efficacious at expelling evil forces and curing diseases (850). Similarly, in the conclusion to his paper Li states that people today no longer fear the Royal Lords but are willing to erect temples for them (859). However, while few would contest the fact that the history of the Royal Lords cult in Taiwan has been marked by the changes described here, the details of when and how such changes occurred are far less certain. As I show later, it is also not at all clear that the changes Li postulates are unique to modern Taiwanese history. One key problem is that while the study of cultural change in twentieth-century Taiwan has received considerable attention during the past decade, the issue of how such changes may have differed from those in China or among overseas Chinese communities in Southeast Asia during the modern era has yet to be adequately addressed. The political histories of each of these regions are of course very different, but we need to learn more about whether such differences may also be reflected in the process of cultural change. For example, while colonial public health policies may have had an impact on popular religion in Taiwan during the early twentieth century, we are yet to fully determine how public health efforts in China and Southeast Asia during the same time period influenced religious activities in these areas. Fortunately, a growing body of research on the modern history of public health and Chinese religion in these areas (see, for example, Benedict 1996; Cheu 1988; Dean 1993a, 1998; Pas 1989; Tan 1990) may provide the basis for more systematic comparisons in the future.

In assessing the overall strengths and weaknesses of the arguments presented here, it is important to recall that while both Tung and Li are leading scholars in the field of Taiwanese religious studies, both scholars share one common belief about the development of the Royal Lords cult in Taiwan: that such changes did not occur in China. For example, Tung maintains that the transformation of the Royal Lords from demons to deities is due to Taiwan's supposedly unique history (Tung 1995: 819). Li adopts a similar position by claiming that the current festive atmosphere at Royal Lords rituals in modern Taiwan contrasts markedly from earlier rites in China (Li 1995: 857, 859). In light of their scholarly credentials, however, it is somewhat surprising that neither Tung nor Li based their conclusions on the growing body of research about popular religion in Zhejiang and Fujian mentioned earlier (see note 5). Li, in particular, has done extensive historical and field research in China, and has published numerous works about the cult of the Royal Lords and its historical antecedents.[12] Nevertheless, he does not systematically compare the Royal Lords cults of Fujian and Taiwan, apart from the brief one-paragraph summary of the contents of Fujian gazetteers.

If Li and Tung had systematically undertaken research on Royal Lords cults in southern China, they would have discovered that many of the changes they describe were hardly unique to Taiwan and that many of their arguments turn out to be highly problematic. Take, for example, the transformation of the Royal Lords from malevolent plague deities to benevolent protective deities. Such a change appears to have accompanied the growth of many cults throughout China, particularly those featuring plague spirits and vengeful ghosts. For example, research on the cult of Marshal Wen has shown that he was originally conceived of as a snake-demon who spread contagious diseases by spitting out poisonous vapors. While these serpentine features were quickly expunged from later versions of Wen's hagiography and iconography, their very presence indicates that at least some worshippers saw him as a potentially dangerous spirit with the power to infect others. It is even possible that the use of the character *wen* (meaning "warm"; also a Chinese surname) instead of *wen* ("fevers" or "epidemics") in Marshal Wen's titles may have represented an attempt to cover up his cult's sinister origins, as well as link it to its cult center in the coastal port of Wenzhou (Zhejiang) (Katz 1995a).

As Wen's cult continued to grow, the nature of his links to epidemics shifted dramatically. While Marshal Wen was originally represented as a demon who could spread epidemics, he eventually came to be worshipped as a deity with the power to prevent them. The exact process by which Wen became transformed from a demon into a deity is not clear, but it appears to have been somewhat similar to what frequently occurs in Taiwan. Field data on a number of cults in Taiwan, including those of some Lords, indicates that the souls of those who die premature or violent deaths, and prove powerful enough to resist attempts at exorcism, are considered to be vicious ghosts. Not all vengeful ghosts are alike, however, as those able to acquire an individual identity and also prove efficacious when approached by worshippers may end up being worshipped as gods.[13] It appears that while Marshal Wen may originally have been represented as a demon who could spread epidemics, he eventually became worshipped as a deity with the power to prevent them.

As for Marshal Wen's festivals, data from the Qing and Republican eras indicate that by at least the eighteenth century, mammoth plague expulsion rites dedicated to Wen and staged in the coastal cities of south China had developed into highly festive events attracting thousands of worshippers, who also attended the colorful temple fairs. Accounts of the festivals of Marshal Wen glowingly describe how the streets were decorated with lanterns, and claim that people partook of only the finest foods. Such sources also emphasize the lavish offerings presented to the gods, and the fine wares people displayed outside their houses when Wen's palanquin passed. Local merchants and other members of the elite competed in presenting elaborate displays of wealth, especially the owners of shops along the procession route. Women would dress in their best clothes and families would show off their finest possessions. Huge feasts were held, dramas were performed, and a festive atmosphere reigned. Sizeable sums of money changed hands during Wen's festivals, with both merchants and peddlers making tidy profits (Katz 1995a).

A similar lack of understanding of China's Royal Lords cults also ends up distorting Tung's and Li's analyses of Taiwan's boat expulsion rituals. Both scholars staunchly maintain that during the Qing Dynasty the Taiwanese people stopped floating wooden boats out to sea and switched to burning boats made of paper or bamboo. However, as I mentioned earlier, rituals involving the expulsion of plague spirits by means of a boat flourished throughout south China since at least the tenth century, and could be performed on behalf of individuals, families, or the entire community. Various forms of these rituals flourished in a number of provinces, including Fujian, Zhejiang, Jiangsu, Jiangxi, Hunan, Sichuan, Guizhou, and Yunnan. The nature of the boat expulsion rites varied from site to site: some communities burned their boats, while others floated them away. Some even floated them away and then set them afire. Local gazetteers are full of official complaints about the excessive expenditures of money occurring at these festivals, but the state appears to have made no attempt to outlaw or reform these events until the twentieth century. Similar boat expulsion rites not only flourished in south China but may also be found throughout Asia in places such as Korea and Tibet, as well as parts of Southeast Asia (see Katz 1995a,b, 1997). Therefore, to claim that one type of boat expulsion ritual may be unique to Taiwan contradicts a great deal of historical and ethnographic evidence to the contrary.

Perhaps the greatest flaw in the work of Tung and Li lies in the way they use their sources, particularly local gazetteers. While such texts are invaluable for scholars studying the regional history of China, they do need to be used with extreme care. One manual for the study of Ming history points out that:

> Even though they were ordinarily compiled by local scholars, gazetteers developed under strong official influence, reflecting the administrative divisions of the empire and the perspectives and preoccupations of the official class. They vary considerably in form, and their quality is uneven. Some appear to have been prepared casually or in haste, while others are comprehensive works of impressive scholarship. (Farmer, et al. 1994: 99)

Furthermore, Hsiao Kung-ch'üan argues convincingly that:

> Some . . . records [in gazetteers] of local conditions, events,and personalities were more painstakingly or competently done than others; but a considerable number are marred by the partiality, dishonesty, or carelessness of contributors. . . . The fact that any single gazetteer was written by a number of persons whose scholarly qualifications were not uniformly high and who frequently executed their assignments with poor coordination and inadequate supervision, points to the possibility of unintentional errors and omissions, even where willful misrepresentation was not practiced. (Hsiao 1960: vii–viii)

The social historian needs to be particularly critical when considering descriptions of local customs presented in gazetteers, as the editors of such works tended to only include data on customs deemed important enough to merit official notice (Hansen 1990: 24). In addition, many gazetteer editors often copied verbatim sections of earlier editions. To his credit, Li notes the existence of some of these problems, as well as the fact that rituals tend to change at a slower pace than the cults they are linked to (Li 1995: 829–831, 842, 857). Nevertheless, he and other scholars researching the history of popular religion in Taiwan still tend to believe that the descriptions of customs in local gazetteers represent actual accounts of contemporary practices. While it is true that these descriptions may partly reflect local customs, it is also readily apparent that they were not based on fieldwork, but most likely on hearsay or recollections of local elders. In addition, descriptions of popular customs are rarely specific about time and place. As we shall see below, one can find a description of a boat expulsion ritual in a county gazetteer without being sure of the exact cult site at which it was performed or whether all cult sites throughout the county featured identical rituals.

The fact that Tung and Li fail to fully comprehend the nature, strengths, and weaknesses of their main primary sources results in distortions of both the data and their arguments. The earliest Taiwan gazetteer to describe the Royal Lords cult and rituals is the Zhuluo county gazetteer of 1717. It states that such rituals, known as "Royal Lords Offerings" (*wangjiao*)[14] were performed by spirit mediums (*wu*) at three-year intervals. According to the text, the Royal Lords boat was originally made of wood and floated out to sea, but this posed a threat to other communities along the coast who would have had to perform offering rituals should the boat land on their shores. As a result, more recent plague boats were built using paper and bamboo, and were burned instead of floated away. A similar passage may be found in the 1720 edition of the Taiwan county gazetteer, although the latter source notes that Taoist priests were in charge of the offering rituals.

In analyzing these two sources, Li places great emphasis on the supposed change in ritual specialists from spirit mediums to Taoist priests (Li 1995: 849–850). In addition, both Tung and Li view the use of paper and bamboo instead of wood to construct Royal Lords boats, as well as the supposed change to boat-burning rituals, as evidence that the process of "indigenization" had occurred (see earlier). Regarding the first point, I would argue that Li has taken the Zhuluo county gazetteer too literally. It is highly unlikely that spirit mediums were ever in charge of Taoist offering rituals, and the use of the exonym/label *wu* in the aforementioned source reflects the contempt of the gazetteer's editors for Taoist specialists. As for the change in materials used to build the Royal Lords boats, it is almost impossible to determine how widespread such a phenomenon was, what "material conditions" may have sparked such a change, and how many cult sites were actually involved. From 1684 to 1731, Zhuluo County effectively comprised the western coast of Taiwan extending from the Xingang River (in present day Jiayi County) to the northern port of Jilong. However, most accounts of customs in local gazetteers tend to reflect practices in areas nearer to the county seat, as such areas were more familiar to local officials. This suggests that the description mentioned above probably refers to Royal Lords cults in and around the town of Zhuluo (present day Jiayi City). However, the number of cults and rituals that actually changed, and their locations, is not indicated and will probably never be known.

The problem of the Royal Lords supposed "indigenization" becomes even more vexing when one consults the 1760 edition of the Fengshan County gazetteer and the 1752 edition of the Taiwan County gazetteer. While the Fengshan gazetteer copies portions of the earlier Zhuluo gazetteer account, it also adds a detailed description of local Royal Lords rituals, noting that Royal Lords Offerings in Fengshan featured the use of *both* wooden and paper Royal Lords boats, the former being floated away and the latter burned at the waterside. The Taiwan County gazetteer is even more important. This work does not copy earlier sources but instead contains a detailed description of local temples dedicated to Royal Lords, the iconography of the deities worshipped therein, and the nature of the plague expulsion rituals performed at such sites. In discussing these rites, the text clearly states that the Royal Lords boat continued to be made of wood, while the statues of the plague spirits (who were *not* the same deities as those worshipped in the Royal Lords temples) and the model implements placed in the boat were made of paper. In his paper, Li quotes a portion of this account out of context (Li 1995: 843), leading the reader to believe that Royal Lords boats at that time were made out of paper when in fact this was not the case.

The most significant aspect of this account is its assertion that local Royal Lords boats were made of wood. While we once again have no way of knowing exactly which cult sites are being referred to, this statement does appear to contradict the passage on Royal Lords rituals in the 1720 edition of the Taiwan county gazetteer compiled just 32 years earlier, which claims that Royal Lords boats were being made of paper and bamboo. Wooden Royal Lords boats continued to be used in Taiwan County, as a work compiled during the early years of the Japanese occupation entitled *Anping xian zaji* also describes the use of such vessels.[15]

Concluding Remarks

What is one to make of this confusing pile of evidence? Does the fact that different local gazetteers contradict each other in describing the growth of various Royal Lords cults mean that such cults did not "indigenize" in Qing-Dynasty Taiwan? Not at all. At the same time, however, I would argue that the data marshaled to date prove insufficient to support any arguments that such a process occurred. This is not to deny that Taiwan's Royal Lords cults may have undergone some changes seldom seen in China, but further data will be necessary before the historical and cultural importance of such changes can be adequately ascertained. One possible change in Taiwan's Royal Lords cults, which both Tung and Li overlook, involves the types of boats used in plague expulsion rituals. Throughout China, including Fujian, such boats were almost always shaped like dragon boats. However, I have yet to find any evidence indicating that dragon boats were used in Taiwan rites, with the possible exception of the White Dragon Abbey in Tainan, which followed the Fuzhou ritual tradition of northern Fujian (Kataoka 1981 (1921): 676; Szonyi 1997). The significance of this phenomenon has yet to be determined.

It is worth noting that while our overall knowledge of the history of China's and Taiwan's Royal Lords cult remains relatively sketchy and incomplete, the evidence collected to date strongly suggests that cults of Royal Lords such as Lord Chi developed in roughly parallel patterns, and that each growth phase appears to have been in part due to a combination of the periodic ravages of contagious diseases as well as socioeconomic developments. It is also clear that the Chinese state could at times be highly critical of Royal Lords cults and their practices and did not consider them "orthodox" enough to be included in the Register of Sacrifices (*sidian*). At the same time though, the state was usually able to tolerate their presence and generally refrained from labeling them as "licentious/illicit sacrifices" (*yinsi*). Only the Japanese colonial government attempted a widespread campaign against Royal Lords (and other) cults, but was largely unsuccessful.[16]

All in all, it appears that arguments in favor of the "indigenization" of Taiwan's Royal Lords cults may have been overly exaggerated. Such cults clearly did change after they spread to Taiwan, but the historical and religious significance of such changes has yet to be adequately understood. The processes of cultic development described above may indicate the presence of "indigenization," but it is also probable that they also reflect processes of change that shaped popular cults throughout China. It seems fair to say that the current state of the field of Taiwanese Studies indicates that any arguments about the "indigenization" of the island's popular religion need to be evaluated in light of what we know about long-term patterns of social and religious history among Han-Chinese communities in China, Taiwan, and Southeast Asia. The exact nature, timing, and significance of cultic development in Taiwan remain topics for future research, and no convincing conclusions will be reached until thorough case studies of specific cults and cult sites have been completed.

Perhaps what the discussion reveals above all is that the use of dichotomies such as "inlandization" or "indigenization" can be rather unproductive, particularly if one steadfastly maintains that these two phenomena are mutually exclusive. As a substitute, one could turn to the "holistic" approach to the study of popular culture discussed by Catherine Bell in an important review article on Chinese (and Taiwanese) popular religion (Bell 1989). As Bell shows, this approach arose in research covering first Western and then Chinese culture as a response to arguments over whether popular religion contributed to cultural unity or diversity. The holistic approach allows for the presence of both unity and diversity, but places even greater emphasis on how ideas, beliefs, and values are created and transmitted. This approach can prove highly useful for scholars studying cultural change at the local level. As Kenneth Dean notes in his new book on the Three in One Religion Sanyi jiao, ". . . perhaps we can imagine a vast variety of locally rooted and constantly changing conceptions of cosmos and individuality, rising out of local and immediate contests of power and metamorphoses of bodies" (Dean 1998: 60).

Adopting such an approach could enable us to study how local cults arose in China, how they spread to Taiwan, and the ways in which they changed (and didn't change) over time. Taiwan should not be viewed as representing a "typical" Chinese culture, and neither should any other region. However, while China's cultural regions have yet to be clearly defined, it appears that Taiwan may belong to an area that includes Fujian and parts of other southeastern coastal provinces mentioned above. Systematic comparisons between popular religion in Taiwan and these parts of China should provide new perspectives on these regions' common points and differences, thus contributing to the study of cultural diversity currently engaging much of the field of Chinese religions.

Notes

1. For a review of the history of such scholarship, see Chang (1995); Nadeau (1996). One should also consult Lin Mei-rong's new and enlarged *Bibliography of Taiwanese Folk Religion* (Lin 1997).
2. For a fascinating appraisal of this debate, see a recent essay by one of the series' editors, Tu Cheng-sheng, Director of the Institute of History and Philology at Academia Sinica (Tu 1998).

3. For detailed discussions of the changes affecting postwar Taiwan and their impacts on local politics, see Bosco (1992); Ch'ü (1988); Ch'ü and Chang (1986); Clart (1995/96); Cohen (1994 [1991]); Harrell and Huang (1994); Hong and Murray (1994); Jordan (1994); Jordan and Overmyer (1986); Li (1982); Lin (1996); Rubinstein (1993, 1994, 1995); Sangren (1996); Sung (1985, 1994a); Ts'ai (1996); Wachman (1994a,b); and Weller (1994, 1999). For an overview of the indigenization movement, see Ch'en (1995). For a discussion of similar problems throughout East Asia, see Befu (1993).
4. The increase from 74 to 447 temples was largely due to better record keeping by the Japanese authorities, although some new temples were also founded during the late nineteenth and early twentieth centuries. At the same time, however, it is important to remember that official statistics concerning temples generally omit the presence of numerous shrines and other small-scale cult sites, which either do not register or are not counted.
5. For more on the history of the Royal Lords and related cults, see Chen and Wu (1993): 149–153; Dean (1986, 1988, 1993a); Katz (1995a,b, 1997); Lagerwey (1987, 1993); Lao and Lü (1993); Lin (1986, 1998); Lin and Peng (1993); Liu (1983); Maejime (1938); Mio (2000); Miyamoto (1988); Shi (1994); Ts'ai (1989); Ts'ai (1994); Xu (1993); and Yen (1994). See also the numerous articles and books on Fujian rituals and ritual dramas published in the journal *Minsu quyi* or its related series on Chinese ritual dramas entitled *Minsu quyi congshu*.
6. For more on the problem of ethnic identity in Taiwan, see Chen et al. (1994); Sangren (1995).
7. See also Tung's remarks on pp. 814 (note 12), 815 (note 13), and 822.
8. See for example Lin and Peng (1993): 349, 359, 362, 363, 367, 368.
9. See Ch'en (1975, 1984, 1987); Li (1975, 1978).
10. The emphasis on the decline of feuds is based on the work of Ch'en Ch'i-nan, who argues that the eventual cooperation between members of different native-place groups in staging local festivals marks one aspect of "localization" (Ch'en 1987).
11. See, for example, Elvin and Liu (1998), as well as Meskill (1979); Shepherd (1993).
12. See, for example, Li (1993a,b,c, 1994).
13. For more on this problem, see Harrell (1994); Katz (1997). The latter work, a book on the Royal Lords I published in Chinese, also contains materials published in Katz (1991, 1994b,c).
14. In terms of their content, Royal Lords' Offerings are essentially the same as Plague Offerings.
15. This source also contains a description of plague expulsion rituals at the White Dragon Abbey (Bailong) An in Tainan City. This temple was linked to temples to the Five Commissioners of Epidemics in Fuzhou, and not surprisingly the rituals performed at the Abbey closely resemble those in Fuzhou.
16. While the governments of late Qing and early Republican China also attempted to suppress local cults (Duara 1991), we know little of the effects of these campaigns on China's Royal Lords cults.

Bibliography

Ahern, Emily M. 1981. "The Thai Ti Kong Festival." In Emily Martin Ahern and Hill Gates, eds., *The Anthropology of Taiwanese Society*. Stanford: Stanford University Press, pp. 397–425.

Anderson, Benedict. 1991. *Imagined Communities. Reflections on the Origin and Spread of Nationalism*. Revised edition. London: Verso.

Baity, Philip. 1975. *Religion in a Chinese Town*. Taipei: Asian Folklore and Social Monograph Series, Number 64.

Befu, Harumi, ed. 1993. *Cultural Nationalism in East Asia: Representation and Identity*. Berkeley: Institute of East Asian Studies.

Bell, Catherine. 1989. "Religion and Chinese Culture: Toward an Assessment of 'Popular Religion'." *History of Religions*, 29: 35–57.

Benedict, Carol. 1996. *Bubonic Plague in Nineteenth-Century China*. Stanford: Stanford University Press.

Bosco, Joseph. 1994 (1992). "The Emergence of a Taiwanese Popular Culture." In Murray A. Rubinstein, ed., *The Other Taiwan*. Armonk: M. E. Sharpe, pp. 392–403. Originally published in *American Journal of Chinese Studies*, 1: 51–64.

Chang Hsün. 1996. "Guangfu hou Taiwan renlei xue Hanren zongjiao yanjiu zhi huigu" [A Review of Anthropological Studies of Han Chinese Religion in Taiwan, 1945–1995]. *Bulletin of the Institute of Ethnology, Academia Sinica*, 81: 163–215.

Ch'en Chao-ying. 1995. "Lun Taiwan de bentuhua yundong: Yige wenhuashi de kaocha" [On Taiwan's Indigenization Movement: An Examination of Cultural History]. In Huang Chün-chieh, ed., *Gaoxiong lishi yu wenhua lunji* [*Collection of Essays on the History and Culture of Gaoxiong*]. Gaoxiong: Chen Zhonghe Weng Cishan Jijinhui. Volume 2, pp. 387–445.

Ch'en Ch'i-nan. 1975. "Qingdai Taiwan Hanren shehui de jianli ji qi jiegou" [The Formation and Structure of Han Chinese Society in Qing-dynasty Taiwan]. M.A. thesis. National Taiwan University.

——. 1984. "Tuzhuhua yu neidihua: Lun Qingdai Hanren shehui defazhan moshi" [Indigenization and Inlandization: A Discussion of Two Forms of Development for Han Chinese Society in Qing-dynasty Taiwan]. *Zhongguo haiyang fazhan shi Junwen ji Collected Essays from the Conference on Chinese Maritime History*] Nankang: Sun Yat-sen Institute of Social Sciences, Academia Sinica, pp. 335–366.

——. 1987. "Lun Qingdai Hanren shehui de zhuanxing" [A Discussion of the Transformation of Han Chinese Society during the Qing Dynasty]. In *Taiwan de chuantong Zhongguo shehui* [*Traditional Chinese Society in Taiwan*]. Taipei: Yunchen congkan. Number 10, pp. 153–182.

Chen Chung-min et al., eds. 1994. *Ethnicity in Taiwan. Social, Historical and Cultural Perspectives*. Nankang: Institute of Ethnology, Academia Sinica.

Chen Geng and Wu Anhui. 1993. *Xiamen minsu* [*The Customs of Amoy*]. Xiamen: Lujiang chubanshe.

Cheu Hock Tong. 1988. *The Nine Emperor Gods. A Study of Chinese Spirit-medium Cults*. Singapore: Time Books International.

Ch'ü Hai-yüan. 1988. "Taiwan diqu minzhong di zongjiao xinyang yu zongjiao taidu" [Popular Religious Beliefs and Attitudes in the Taiwan Region]. In Yang Kuo-shu and Ch'ü Hai-yüan, eds., *Bianqian zhong de Taiwan shehui* [*Taiwan Society in Transition*]. Nankang: Institute of Ethnology, Academia Sinica.

Ch'ü Hai-yüan and Chang Ying-hua, eds. 1986. *Taiwan shehui yu wen- hua bianqian* [*Social and Cultural Change in Taiwan*]. Nankang: Institute of Ethnology, Academia Sinica.

Chongxiu Taiwan sheng tongzhi. 1992. Nantou: Taiwan sheng wenxian weiyuan hui.

Clart, Philip. 1995/96. "Sects, Cults and Popular Religion: Aspects of Religious Change in Postwar Taiwan." *B.C. Asian Review*, 9: 120–163.

Cohen, Myron. 1994 (1991). "Being Chinese: The Peripheralization of Traditional Identity." In Tu Wei-ming, ed., *The Living Tree. The Changing Meaning of Being Chinese Today*. Stanford: Stanford University Press. First published in *Daedalus*, 120.2.

Dean, Kenneth. 1986. "Field Notes on Two Taoist *Jiao* Observed in Zhangzhou in December, 1985." *Cahiers d'Extrême Asie*, 2: 191–209.

——. 1988. "Manuscripts from Fujian." *Cahiers d'Extrême Asie*, 4: 217–226.

——. 1993a. *Taoist Ritual and Popular Cults of Southeast China*. Princeton: Princeton University Press.

——. 1993b. "Conferences of the Gods. Popular Cults Across the Taiwan Straits." Paper presented at Conference on Economy and Society in Southeastern China. Cornell University, October 2–3.

——. 1998. *Lord of the Three in One. The Spread of a Cult in Southeast China*. Princeton: Princeton University Press.

Duara, Prasenjit. 1991. "Knowledge and Power in the Discourse of Modernity: The Campaigns against Popular Religion in Early Twentieth-Century China." *Journal of Asian Studies*, 50.1: 67–83.

Elvin, Mark and Liu Ts'ui-jung, eds. 1998. *Sediments of Time Environment and Society in Chinese History*. Cambridge: Cambridge University Press.

Fan Yen-ch'iu. 1994. "Riju qianqi Taiwan zhi gonggong weisheng -yi fangyi wei zhongxin zhi yanjiu (1895–1920)" [Public Health during the Early Japanese Occupation of Taiwan—A Study Centering on Plague Prevention (1895–1920)]. M.A. thesis. National Taiwan Normal University.

Farmer, Edward L. et al., comp. 1994. *Ming History. An Introductory Guide to Research*. Minneapolis: Ming Studies Research Series, number 3.

Gold, Thomas B. 1986. *State and Society in the Taiwan Miracle*. Armonk, N.Y.: M. E. Sharpe.

——. 1994. "Civil Society and Taiwan's Quest for Identity." In Stevan Harrell and Huang Chün-chieh, eds., *Cultural Change in Postwar Taiwan*. Boulder: Westview Press, pp. 47–68.

Hansen, Valerie. 1990. *Changing Gods in Medieval China, 1127–1276*. Princeton: Princeton University Press.

Harrell, C. Stevan. 1974. "When a Ghost Becomes a God." In Arthur P. Wolf, ed., *Religion and Ritual in Chinese Society*. Stanford: Stanford University Press, pp. 193–206.

Harrell, C. Stevan and Huang Chün-chieh, eds. 1994. *Cultural Change in Postwar Taiwan*. Boulder: Westview Press.

Holcombe, Charles. 1995. "Re-imagining China: The Chinese Identity Crisis at the Start of the Southern Dynasties Period." *Journal of the American Oriental Society*, 115.1: 1–14.

Hong Keelung and Stephen O. Murray. 1994. *Taiwanese Culture, Taiwanese Society: A Critical Review of Social Science Research Done on Taiwan*. Lanham: University Press of America.

Hsiao Kung-ch'üan. 1960. *Rural China Imperial Control in the Nineteenth Century*. Seattle and London: University of Washington Press.

Jordan, David K. 1976. "The Jiaw (Jiao) of Shigaang (Xigang): An Essay in Folk Interpretation." *Asian Folklore Studies*, 35: 81–107.

——.1994. "Changes in Postwar Taiwan and their Impact on the Popular Practice of Religion." In Harrell and Huang, eds., *Cultural Change in Postwar Taiwan*, pp. 137–160.

Jordan, David and Daniel Overmyer. 1986. *The Flying Phoenix. Aspects of Chinese Sectarianism in Taiwan*. Princeton: Princeton University Press.

Kataoka Iwao. 1981 (1921). *Taiwan fengsu zhi* [*Gazetteer of Taiwan Customs*]. Trans. Ch'en Chin-t'ien and Feng Tso-min. Taipei: Ta-li Publishing Company.

Katz, Paul. 1987. "Demons or Deities?—The *Wangye* of Taiwan." *Asian Folklore Studies*, 46.2: 197–215.

——. 1991. "Pingdong xian Donggang zhen de yingwang jidian: Taiwan wenshen yu wangye xinyang zhi fenxi" [The Welcoming the Lords Festival of East Haven, Pingdong County: A Study of Taiwan's Plague Gods and Lords Beliefs]. *Bulletin of the Institute of Ethnology, Academia Sinica*, 70: 95–211.

——. 1994a. "Welcoming the Lords and the Pacification of Plagues. The Relationship between Taoism and Local Cults." Paper presented at the Association of Asian Studies Annual Meeting. Boston, March 24–27.

——. 1994b. "Commerce, Marriage and Ritual: Elite Strategies in Tung-kang During the Twentieth Century." In Chuang Ying-chang and Pan Ying-hai, eds., *Taiwan yu Fujian shehui wenhua yanjiu lunwenji* [*Collection of Essays on Society and Culture in Taiwan and Fujian*]. Nankang: Institute of Ethnology, Academia Sinica, pp. 127–165.

——. 1994c. "Zhanhou wangye xinyang de yanbian—Yi Donggang Donglong Gong ji Taibei Sanwang Fu wei li" [Changes in Royal Lords Cults in Postwar Taiwan—A Case Study of the Temple of Eastern Prosperity in East Haven and the Hall of the Three Lords in Taibei]. In Sung Kuang-yü, ed., *Taiwan jingyan (2)—Shehui wenhua pian* [*The Taiwan Experience (2)—Society and Culture*]. Taipei: Tung-ta Publishing Company, pp. 161–174.

——. 1997. *Taiwan de wangye xinyang* [*The Cult of the Royal Lords in Taiwan*]. Taipei: Shangding Publishing Company.

Lagerwey, John. 1987. *Taoist Ritual in Chinese Society and History*. New York: Macmillan Publishing Company.

—— (Lao Gewen). 1993. "Fujian sheng nanbu xiancun Daojiao chutan" [Preliminary Investigations into the State of Daoism in Southern Fujian]. *Dongfang zongjiao yanjiu*, 3: 147–168.

Lao Gewen and Lü Ts'ui-k'uan. 1993. "Zhejiang sheng Cangnan diqu de Daojiao wenhua" [Taoist Culture in the Cangnan Region of Zhejiang]. *Dongfang zongjiao yanjiu*, 3: 171–198.

Li Feng-mao. 1993a. "Donggang wangchuan, howen, yu songwang xisu zhi yanjiu" [A Study of the Plague Boat, the Pacification of Plagues Ritual, and the Sending off the Lords Rituals at East Haven]. *Dongfang zongjiao yanjiu*, 3: 229–265.

——. 1993b. "Taiwan Donggang pingan jidian de wangye raojing yu hojing pingan" [The Procession of the Lords and the Concept of Local Tranquility as Seen in in East Haven, Taiwan]. *Minsu quyi*, 85: 273–323.

——. 1993c. *Donggang wangchuanji* [*The Plague Boat Sacrifice at East Haven*]. Pingdong: Pingdong County Government.

——. 1994. "Xingwen yu songwen—wenshen xinyang yu zhuyi yishi de yiyi" [Spreading and Sending off Epidemics—The Significance of Plague God Beliefs and Plague Expulsion Rituals]. In *Proceedings of the International Conference on Popular Beliefs and Chinese Culture*. Taipei: Center for Chinese Studies. Volume I, pp. 373–422.

——. 1995. "Taiwan songwen, gaiyun xisu de neidi hua yu bendihua." In *Diyi jie Taiwan bentu wenhua xueshu yantao hui lunwenji* [*Proceedings of the First Academic Conference on Taiwan's Indigenous Culture*]. Taipei: National Teacher's University, pp. 829–861.

Li Kuo-ch'i. 1975. "Qingji Taiwan de zhengzhi jindai shi—Kai-shan, fufan yu jiasheng (1875–1894)" [The Modern Political History of Taiwan during the Qing—Settlement, Revival, and Augmentation]. *Zhonghua wenhua fuxing yuekan*, 8.12: 4–16.

——. 1978. "Qingdai Taiwan shehui de zhuanxing" [The Transformation of Taiwanese Society during the Qing]. *Zhonghua xuebao*, 5.3: 131–159.

Li Yih-yüan. 1982. "Taiwan minsu xinyang de fazhan qushi" [Trends in the Development of Popular Beliefs in Taiwan]. In *Minjian xinyang yu shehui yantao hui* [*Conference on Popular Beliefs and Society in Taiwan*]. Taichung: Tung-hai University, pp. 89–101.

Lin Fu-shih. 1986. "Shishi Shuihudi Qinjian zhong de 'li' yu 'ding-sha'" [A Preliminary Explanation of the Terms "li" and "dingsha" Found in Qin Dynasty Bamboo Slips]. *Shiyuan*, 15: 2–38.

——. 1998. "The Cult of Jiang Ziwen in Medieval China." *Cahiers d'Extrême Asie*, 10: 335–57.

Lin Guoping and Peng Wenyu. 1993. *Fujian minjian xinyang* [*Popular Beliefs in Fujian*]. Fuzhou: Fujian renmin chuban she.

Lin Mei-rong. 1996. *Taiwan wenhua yu lishi de chongjian* [*The Reconstruction of Taiwan's Culture and History*]. Taipei: Qianwei Publishing Company.

——. ed. 1997. *Taiwan minjian xinyang yanjiu shumu (zengding ban)* [*A Bibliography of Taiwanese Folk Religion (Enlarged Edition)*]. Nankang: Institute of Ethnology, Academia Sinica.

Lin Yü-t'i. 1987. *Taiwan jiaoyu mianmu sishi nian* [*The Faces of Taiwan's Education over the Past Forty Years*]. Taipei: Ziliwanbao, Cultural Division.

Liu Chih-wan. 1983. *Taiwan minjian xinyang lunji* [*Collected Essays on Taiwan's Popular Beliefs*]. Taipei: Lien-ching Publishing Company.

Maejime Shinji. 1938. "Taiwan no onyakugami, ōya, to sōō no fūshū ni tsuite" [On Taiwan's Plague Gods, the Royal Lords, and the Custom of Sending off the Plague Boat]. *Minzokugaku kenkyū*, 4.4: 25–66.

Meskill, Johanna M. 1979. *A Chinese Pioneer Family*. Princeton: Princeton University Press.

Mio Yuko. 2000. "Taiwan wangye xinyang de fazhan: Taiwan yu Dalulishi he shikuang de bijiao" [The Development of the Cult of the Royal Lords on Taiwan: A Comparison of the History and Current State of the Cult in Taiwan and the Mainland]. In Hsü Cheng-kuang and Lin Mei-rong, eds., *Renlei xue zai Taiwande fazhan. Jingyan yanjiu pian* [*The Development of Anthropology on Taiwan. Papers on Research Experiences*]. Nankang: Institute of Ethnology, Academia Sinica, pp. 31–67.

Miyamoto Nobuto. 1988. *Nihon tōchi jidai Taiwan ni okeru jibyō seiri mondai* [*The Problem of "Temple Rectification" during the Japanese Rule over Taiwan*]. Nara: Tenrikyō dōyūsha.

Nadeau, Randall L. 1996. "Chinese Religion: The State of the Field in Chinese Scholarship." Paper presented at the American Academy of Religion Annual Meeting. New Orleans, November 23–26.

Pas, Julian F. ed. 1989. *The Turning of the Tide. Religion in China Today*. Hong Kong: Oxford University Press.

Rohsenow, Hill Gates. 1973. "Prosperity Settlement: The Politics of Paipai in Taipei, Taiwan." Ph.D. thesis. University of Michigan.

Rubinstein, Murray A. 1993. "Fujian's Gods/Taiwan's Gods. Reconstructing the Minnan Religious Matrix." Paper presented at Conference on Economy and Society in Southeastern China. Cornell University, October 2–3.

——. ed. 1994. *The Other Taiwan*. Armonk: M. E. Sharpe.

——. 1995. "Statement Formation and Institutional Conflict in the Mazu Cult" Temples, Temple-created Media, and Temple Rivalry in Contemporary Taiwan." In Chou Tsung-hsien, ed., *The International Academic Conference on the History of Taiwan: Society, Economics, and Colonization*. Taipei: Kuo—shih kuan, pp. 189–224.

——. 1999. *Taiwan. A New History*. Armonk: M.E. Sharpe.

Sangren, P. Steven. 1996. "Anthropology and Identity Politics in Taiwan: The Relevance of Local Religion." Fairbank Center Working Papers. Number 15.

Shepherd, John R. 1993. *Statecraft and Political Economy on the Taiwan Frontier, 1600–1800*. Stanford: Stanford University Press.

Shi Yilong. 1994. "Tongan Lücu cun de wangye xinyang" [The Royal Lords Cult of Lücu Village in Tongan]. In Chuang Ying-chang and Pan Ying-hai, eds., *Taiwan yu Fujian shehui wenhua yanjiulunwenji* [*Collection of Essays on Society and Culture in Taiwan and Fujian*]. Nankang: Institute of Ethnology, Academia Sinica, pp. 183–212.

Sung Kuang-yü. 1985. "Taiwan minjian xinyang de fazhan qushi" [Trends in the Development of Popular Beliefs in Taiwan]. *Hanxue yanjiu* [*Chinese Studies*], 3.1: 220–226.

——. 1994a. "Shilun sishi nian lai Taiwan zongjiao de fazhan" [A Preliminary Discussion of the Development of Religion in Taiwan during the Past 40 Years]. In idem., ed., *Taiwan jingyan (2)—Shehui wenhua pian* [*The Taiwan Experience (2)—Society and Culture*]. Taipei: Tung-ta Publishing Company, pp. 175–224.

——. 1994b. "Guanyu shanshu de yanjiu ji qi zhanwang" [On Research About Morality Books and its Future Prospects]. *Xin shixue*, 5.4 (December 1994): 163–191.

Szonyi, Mike. 1997. "The Illusion of Standardizing the Gods: The Cult of the Five Emperors in Late Imperial China." *The Journal of Asian Studies*, 56.1: 113–135.

Tan Chee-Beng. 1990. "Chinese Religion and Local Chinese Communities in Malaysia." In idem., ed., *The Preservation and Adaptation of Tradition: Studies of Chinese Religious Expression in Southeast Asia. Contributions to Southeast Asian Ethnography*, 9: 5–27.

Ts'ai Chin-t'ang. 1994. *Nihon teikoku shūgi ka Taiwan no shūkyō seisaku* [*Religious Policies during the Period of Japanese Rule over Taiwan*]. Tokyo: Dohsei Publishing Company.

Ts'ai Hsiang-hui. 1989. *Taiwan de wangye yu Mazu* [*Taiwan's Royal Lords and Mazu Cults*]. Taipei: Taiyuan Publishing Company.

Ts'ai Tu-chien. 1996. "Dui 1980 niandai Taiwan minzu rentong de wenhua fenxi" [A Cultural Analysis of Taiwanese Ethnic Identity during the 1980s]. In Chang Yen-hsien et al., eds., *Taiwan jin bainian lishi lunwenji* [*Proceedings of the Conference on Taiwan History during the Past 100 Years*]. Taipei: Wu San-lien Foundation, pp. 303–330.

Tu Cheng-sheng. 1998. "Lishi jiaoyu yu guojia rentong—Taiwan lishi jiaokeshu fengbo de fenxi" [History Education and National Identity—An Analysis of the Controversy over Taiwan's History Textbooks]. In *Taiwan xin, Taiwan hun* [*Taiwan Heart, Taiwan Soul*]. Kaohsiung: Ho-pan Publishing Company.

Tung Fang-yüan. 1995. "Taiwan minjian xinyang zhi zhengshi" [A Correct View of Taiwan's Popular Beliefs]. In *Diyi jie Taiwan bentu wenhua xueshu yantao hui lunwen ji* [*Proceedings of the First Academic Conference on Taiwan's Indigenous Culture*]. Taipei: National Teacher's University, pp. 809–828.

Wachman, Alan M. 1994a. *Taiwan: National Identity and Democratization*. Armonk: M. E. Sharpe.

——. 1994b. "Competing Identities in Taiwan." In Rubinstein, ed., *The Other Taiwan*, pp. 17–80.

Watson, James. 1985. "Standardizing the Gods: The Promotion of T'ien Hou ('Empress of Heaven') Along the South China Coast, 960–1960." In David Johnson et al., eds., *Popular Culture in Late Imperial China*. Berkeley: University of California Press, pp. 292–324.

Weller, Robert P. 1987. "The Politics of Ritual Disguise: Repression and Response in Taiwanese Religion." *Modern China*, 13.1: 17–39.

——. 1994. *Resistance, Chaos and Control in China: Taiping Rebels, Taiwanese Ghosts and Tiananmen*. Seattle: University of Washington Press.

——. 1996. "Matricidal Magistrates and Gambling Gods: Weak States and Strong Spirits in China." In Meir Shahar and Robert Weller, eds., *Unruly Gods. Divinity and Society in China*. Honolulu: University of Hawaii Press, pp. 250–268.

——. 1999. "Identity and Social Change in Taiwanese Religion." In Rubinstein, ed., *Taiwan. A New History*, pp. 339–365.

Wilson, Richard W. 1970. *Learning to be Chinese: The Political Sinicization of Children in Taiwan*. Cambridge, MA: MIT Press.

Winckler, Edwin A. 1994. "Cultural Policy in Postwar Taiwan." In Harrell and Huang, eds., *Cultural Change in Postwar Taiwan*, pp. 22–46.

Xu Xiaowang. 1993. *Fujian minjian xinyang yuanliu* [*The Origins and Development of Popular Beliefs in Fujian*]. Fuzhou: Fujian jiaoyu chuban she.

Yen Fang-tzu. 1994. *Lugang wangye xinyang de fazhan xingtai* [*The Pattern of Development of Lukang's Royal Lords Cults*]. M.A. thesis. National Tsing-hua University.

Yü Kuang-hung. 1983. "Taiwan diqu minjian zongjiao de fazhan" [The Development of Popular Religion in Taiwan]. *Bulletin of the Institute of Ethnology, Academia Sinica*, 53: 67–105.

Chapter 7

"Medium/Message" in Taiwan's Mazu-Cult Centers: Using "Time, Space, and Word" to Foster Island-Wide Spiritual Consciousness and Local, Regional, and National Forms of Institutional Identity

Murray A. Rubinstein

Introduction

In this chapter, I study the production of "medium/message" by the major Mazu temples of Taiwan. I do so to suggest how this medium/message dyad (or rather this potential equation—for the medium can become the message), adopted and adapted from Marshall McCluan's famous schema, has been made use of by each of three major regional Mazu temples. These three temples—the Hsinkang Fon-t'ien Kung, the Peikang Ch'ao-t'ien Kung, and the Tachia Chen-lan Kung—make use of what I see as a trinity of linked medium—"time," "space," and "word" and use the mediums-as-messages thus produced as instruments or weapons (if one may consider advertisements or publicity devices as weapons) when they interact with or set themselves up against (in the public's mind) each other.

To study interaction among temples[1] is also to study the production and the nature of identity—as manifested in local, regional, provincial, and trans-provincial levels.[2] Identity and its production is a related theme that runs through this chapter, and it is, of course, one that relates to the larger subject of this volume. P. Steven Sangren has shown us in his ambitious, carefully argued, and powerful new essay,[3] that temple rivalry—and the conflict that it produces—is closely tied to Minnan ethnic identity or, more precisely, to multileveled forms of identity production. One can discern these multiple levels of identity and formation/ production as well as patterns of competition when one examines the major temples. Thus, even as I examine "medium and message," I also suggest how the process of medium/message formation and the larger processes of temple-to-temple interaction can be linked to the production of different levels and types of identities—identities that multifaceted and multileveled in their nature—identities that are both spiritual and ethnic, as well as local, regional, provincial (that is, Taiwan-centered) and inter-provincial (here meaning Taiwan and Fukien).

The scholarly foundation for this chapter is found in Sangren's articles and in the work of other scholars who have written about Mazu: These scholars include James Watson, Huang Mei-yin, and Chang Hsun.[4] My own conceptualization of the processes of interaction and rivalry among Taiwan's major Mazu temples and the related issue of the production of ethnic identity, began with a close reading of these scholars' presentations of the evidence and their differing sets of interpretations of that evidence.

This chapter is organized into five sections. The first section examines the background of the goddess Mazu and the development of the Mazu cult in Fukien Province and in Taiwan. The second, third, and fourth sections examine specific components of the medium/message dyad—time, space, word—as means of understanding how three key temples—the Hsin-kang Fon-t'ien Kung, the Pei-kong Ch'ao-t'ien Kung, and the Ta-chia Chen-lan Kung—describe and define themselves for the general public by creating images of who they are and describing what their respective temples and temple community is like. In the fifth and concluding section, I suggest how the set of medium/message dyads produced by each temple are used to foster both various levels of identity and also to allow the temples to compete for the attention of the millions of the island's believers in the goddess Mazu and the Mazu Cult, specifically in Fukien and Taiwan.

Time

The Development of Mazu and her Cult

One way to begin to understand the spirit of Minnan China's popular religiosity is to study the cult of the goddess Mazu—or T'ien-shang Sheng-mu (the Heavenly Mother, to call her by the title bestowed upon her by the Ch'ing rulers) the goddess of seafarers. Populations on both sides of the Taiwan Strait, and along the mainland coast to the north and the south of that treacherous body of water, recognize her power. She has become, over the course of the last millennium, the reigning deity of those who ply China's coastal and offshore waters. Her temples dot the landscapes of provinces from Shantung to Kuangtung and she is recognized to be one of the two or three major figures in the Taiwanese popular pantheon.[5]

Mazu was the unmarried daughter of a family named Lin. The Lin were descendants of a family of officials, and pious Buddhists, who had moved from the Minnan entrepot of Ch'uanchou to the P'ut'ien County town of Kang-li. Here, they became dominant members of a major P'ut'ien County clan. They were a family, like many others in the area, that earned their living from the sea. The literature that one finds in most Mazu temples in Fukien and on Taiwan tells us that this woman, Lin Mo-niang, experienced a birth highlighted by signs and wonders, in 960 A.D. She then went on, this text tells us, to lead the life of an unmarried woman, an unusual circumstance for a woman born into a family of modest means in traditional China. It was, as things turned out, a life that was fulfilling nonetheless for throughout she demonstrated her talent for absorbing knowledge, her possession of special skills, and of her ability to perform wondrous acts and miracles. Furthermore, Lin Mo-niang did not die an ordinary death, but ascended to heaven from Meichou, a island located offshore of her Kangli home, as the final act of her brief, but eventful, existence. While she was alive, those around her in the seaside community that the Lin family made their home, believed that she was a woman of power—either a witch, a shaman, or something more than mortal.[6] Thus, her ascension was something expected.

In the years following her ascension, the legends about her deeds grew and these tales were augmented by accounts of those who had seen her—in the form of a spectral light—guiding ships to safe harbors and protecting the fishermen and the merchants who plied the dangerous waters of the Taiwan strait. She became thought of as a goddess by the local people, but was only one of many such local deities. Only when she saved a government official floundering at sea did the Sung Emperor of that time, in the twelfth century recognize her power and grant her an official title.[7]

In the decades and the centuries that followed, later Sung rulers granted her other titles, and this practice was continued by the rulers of the Yuan, the Ming, and the Ch'ing dynasties. As her influence as a goddess grew, so too, did her cult. Temples that made her the center of worship grew and spread to other parts of Fukien, then into the other provinces of coastal China,[8] to P'enghu (the Pescadores) to the Chinese communities in the Nanyang, and finally to the island of Taiwan.[9]

The development of rivalry is apparent from the very beginnings of the Mazu cult in Fukien among those temples claiming to be the *tsu-miao* (the ancestral temple). The *tsu-miao* claims precedence and authority over the other temples in the region. Perhaps the most important of the *tsu-miao* is the temple devoted to the goddess that is located on Meichou island a few miles offshore of her home village on the P'ut'ien county coast. It was here, according to the mythic biography that one finds in most temple literature, that Lin Mo-niang ascended to heaven. This Tien-hou Kung has made claims of supremacy within the cult since the Sung Dynasty. However, such claims have been contested by Mazu temples in Ch'uanchou and in T'ungan county, an agricultural area to the east of the metropolis of Changchou and north and west of the present-day metropolis of Hsiiamen. T'ungan is also the home of one of the *tsu-miao* of Wu T'ao, the man who is now worshiped as the physician-god Pao-sheng Ta-ti.[10] In this area temples to Mazu and Pao-sheng Ta-ti are found near or next to each other. The traditional literature one finds in this area link Mazu and Pao-sheng Ta-ti are what is described as a romance of Minnan gods.[11]

Both Ch'uanchou and the T'ungan areas were major centers of immigration to Taiwan during the Dutch, Ming/Cheng, and Ch'ing eras and the traditions of religion and temple rivalry were carried with them. The major temples to the goddess that developed on Taiwan after the 1600s followed the well established patterns and engage in struggles with each other, basing their claims for authority on being the first temple in a given county or region or being the local or regional temple with the direct and very early ties to the *tzu-miao* on Meichou Island.

The two temples in the Luerhmen area both claim descent from Meichou. The Anping temple, near this area in Tainan County also claims descent from the temple on that small island, though through the Ma-kung temple. Further north, the older of the Lugong T'ien-hou Kung links itself to the Meizhou temple while the younger Mazu temple, a few blocks away, claims descent from the Ch'uan-chou T'ien-hou Kung.

These temples were founded by people from different counties in Fukien. The temples serve as symbols and as centers for descendents of the members of the temples. The three temples in this essay fit this pattern. The Fon-t'ien Kung in Hsinkong and the Ch'ao-t'ien Kung in Peikong each claim claim to have been founded in the seventeenth century by individuals from Meichou, but their respective accounts of the events differ. And the Tachia temple, located to the north and west of Taichung, which was a client to both of these southern temples—at different times, began later than either of them.

The results of the process of growth have been very impressive. In her useful overview of the Mazu, Chang Hsun has estimated that there are over 300 temples devoted to the goddess in Taiwan.[12] To this very day these major temples continue to build networks with smaller temples, temples that send delegations of pilgrims to their respective mother one or more times a year. These major temples also continue to engage in an ongoing series of conflicts with each other as they try to win the devotion of the people of Taiwan. Rivalry and conflict between these temples and the forms that the conflict takes are explored in the following.

Hsinkong, Peikong, and Tachia: The Medium/Message Dyad and the Production of Public Images of the Major Mazu Temples

The three Taiwanese temples that we focus on—the Hsinkong Fon-t'ien Kung, the Peikong Ch'ao-t'ien Kung, and the Tachia Chen-lan Kung—can be seen as major players in the world of Taiwan's Mazu temples. In this section, I introduce a schema—a medium/message dyad—that is used to suggest the nature of the instruments—or weapons—of persuasion that these temples make use of when they engage in spiritual/social conflict with their major rivals. The medium/message dyad is a variation of the Marshall McLuhan idea that the medium is the message. I would suggest that there are distinct mediums that generate different sets of messages. One can see such medium/message dyads as instruments that a temple committee utilizes in presenting itself to its public and in defining its identity as a religious entity. I will spell out three such medium/message dyads—dyads I label time—history, space—the temple itself, and word—the temples' publication efforts.

Temple Histories and the Reconstruction of the Past

The first medium message dyad to be discussed is encapsulated by the term "time." Here "time" means "history." The committees that run the major Mazu temples in Taiwan, devote considerable space in their large-scale publications to the history of their temples. In each of these histories are found narratives of that temple's development. The histories in these temple-produced volumes begin at an earlier period in time than the actual founding of the temple itself. These accounts tell of the life of Mazu and then trace the growth of her cult in Fukien and along the

China coast. As they do so, they discuss the various titles the goddess is given as the centuries roll by. With this mythic/historical context in place, the authors of the histories then shift to questions of temple origin and development. In laying out the origins of a temple such as those at Hsinkong or the one at nearby Peikong, the authors stress the links to that temple's *tsu-miao*—usually Meichou, and if there is a particularly vivid legend related to a temple's Meichou connection this too will be included in the account. The narratives then cover the major events in the temple's development. In the Peikong/Hsinkong area—once the town of Penkong natural disasters such as floods were common and changed the face of the land and the buildings. As we shall see, the histories of these two temples are filled with accounts of its impact on the course of the river system itself.

Controversy is never far from the surface in the actual histories of the temples and in the myth-laden histories that the major temples publish. Such controversies are found in these histories, with the temple committees giving credence to any controversy, as might be expected. The history of each of these temples contains accounts of the way the goddess has blessed that particular temple. Miracles are relatively common occurrences and these accounts are intended to show the reader—and potential believer—that the goddess has blessed that temple with her spiritual power.

These official "mythic" histories are rich sources of information for journalists and popular writers, who are aware of the Taiwan public's demand for works of history and magic. Thus, the accounts we find in ostensibly secular and objective works parallel these accounts found in temple publications and accept as valid the accounts of miraculous intervention. The journalists and writers know their public and give them what they seem to want. Let us now examine the histories of the three temples, seeing them as mediums, and then move on to discover what messages such histories contain.

The Hsinkong Fon-t'ien Kung. The first of the three Mazu temples to be discussed is the temple that claims to be the first of the Mazu temples built on the island of Taiwan. This is the Hsinkong Fon-t'ien Kung. The temple is located in the small town of Hsinkong, in an area on the southwestern coastal plain that lies west of the county seat of Chiayi and the national highway and twelve kilometers east of Peikong, the site of its famous rival, the Chao-t'ien Kung. This temple was constructed in what was then called the port of Penkong.

The port of Penkong was one of the first areas settled by Minnan Chinese, and this settlement took place even before the Dutch era, dating to the period when the island was under the nominal control of Cheng Ch'en-kong's father, Cheng Chih-lung. Disastrous floods changed the very face of the landscape and wiped out some areas, while separating others from the very town they had been part of. This difficult and complex history helped define the development of the Mazu cult in this area.

The administration of the Hsinkong Fon-t'ien Kung, claims that their temple is the oldest Mazu temple in Taiwan. They state that it was founded in 1622, the period when Ch'eng Chi-lung, claimed power over the island. During this period, on the eve of the Dutch takeover of the area around modern Tainan county, Ch'eng was said to have settled thousands of his troops and other Minnan people in present-day Chiayi county. The coastal town of Penkong was one of those he helped establish.

Many of these early settlers were fishermen, whose lives were dependent upon the treacherous waters of the Taiwan Strait: Their patron goddess and protector was Mazu. It was in 1622 that one of the residents of the new town, a man named Lin, invited Mazu to come to the new town and protect its people. While onboard ship, then docked in Penkong, this man prayed to Mazu and requested that she allow the *shen-hsiang*, which had come from the Meichou *tsu-miao* and that, thus, contained the essential spirit of the goddess, her *ling*, to stay in the town. The goddess, speaking to Lin through the statue, agreed to move onto the land and to permanently serve as the protector of the island of Taiwan and its Minnan inhabitants. A simple temple was constructed and this *shen-hsiang* became the focal point of worship even as other relics and the *shen-hsiang* of other gods were put in place. In the decades that followed, the temple, though small became an important site in the town.

In 1700, sixteen years after the Ch'ing wrested control of the island from the Ch'eng family, a larger and more substantial temple was constructed. This temple was formally named the Chu-lo T'ien-fei Kung. One may assume that the establishment of another center of Mazu worship in the area, this one organized by a monk from Fukien, prompted the administrators of the temple to take this major step. That other center became over time, the Ch'au-t'ien Kung—the famous and more powerful rival of the older temple.

The Chu-lo T'ien-fei Kung collapsed, however, and had to be reconstructed. At this time, Mazu was given yet another new title by the Ch'ing, T'ien-hou, and this raised her status even higher. State support of the cult of the goddess gave the temple administrators the opportunity to raise the status of the reconstructed temple and they did so, calling it the T'ien-hou Kung.

The goddess' newfound glory proved valuable nine years later when the temple collapsed again and had to be rebuilt. The people of the town gave generously to support this effort and a new and more glorious structure was built on the old site. This temple was also given the name of the Penkong T'ien-fei Kung. This newly built temple was expanded further in the years that followed and was rebuilt once again before the disastrous year of 1799.

In 1799, the town of Penkong suffered a flood that changed the course of the major river in the area and destroyed the temple. The *shen-hsiang* revered because of the temple's long history and its connection to Meichou were saved, however, and were moved to Penkong Chieh in the town of Hsinkong.

It was at this moment that another near miracle took place. A military official who lived in the area, Wang Teh-lu built a temple to Mazu as a way of thanking her for his promotions and his worldly success. He also petitioned the emperor to call the temple the Fon-t'ien Kung. Thus, this famous old temple was reborn with a different name but with the same powerful *shen-hsiang* dominating the place of honor in the center of the sanctuary.

In the late nineteenth century nature again took its toll. This time, during a period of years in the Kuansu Era, a succession of earthquakes struck the town. As a result, parts of the temple were destroyed and only sections of what had been destroyed were rebuilt. When residents of the town learned that the entire structure had to be rebuilt they recognized how much a part of Penkong/Hsinkong's identity the temple had become. It was then that they raised the 100,000 yen—for Taiwan was now under the Japanese—that was needed. The work was completed in 1918. Ten years later, the Showa Emperor declared that the temple was a national treasure and had to be prevented from further destruction.

Further restoration and expansion of the temple took a renewed significance after Retrocession. In 1951 two new buildings were constructed in the rear of the main structure and nine years after that, in 1960, the East Hall and the West Hall were rebuilt with new stories added to each hall. In 1972 the four hundred and fiftieth anniversary of the temple was celebrated and Shen-shih Chi, a man who helped found the temple was honored. A new five-tier pagoda was built in honor of the event and as a symbol of the role that this temple had played as a center for those who believed in the great Goddess, T'ien-shang Sheng-mu.[13]

The Peikong Chao-t'ien Kung. The Chao-t'ien Kung, like the Fon-t'ien Kung, was founded in the city of Penkong. In 1694, a Ch'an monk of the Lin Chi School named Shu Pi brought a *shen-hsiang* from Meichou to the small Chiayi County port. The goddess instructed the monk that she wanted her image to stay permanently in the port. Penkong had been settled by Minnan Chinese from Ch'uanchou and Changchou who were faithful believers in Mazu and they invited the monk to take charge of matters related to the care of the *shen-hsiang* and established a proper site where they could worship her.[14]

The structure that was built was a modest one and remained so until 1720. That year a temple was constructed. Ten years later the temple was expanded. However, it was not until 1774 that it was given its name, the Chao-t'ien Kung. Over the course of the next century, it was expanded further and began to serve as both the center of the town of Peikong and the center of the expansion-oriented Mazu cult. In 1894 disaster struck. The town was leveled by a fire and the front hall of the temple was destroyed. The next year saw a new disaster—Taiwan was taken over by Japan as spoils of war. The Taiwanese resisted the occupation and many fought in a guerilla war that dragged on for years. It was only in 1904 that things had settled down sufficiently for people to rebuild the temple. However, nature showed its cruel side once again the next year: an earth quake struck Chiayi and the Chau-t'ien Kung was severely damaged. Funds amounting to 79,000 yen were collected as donations poured in from all parts of the island. With the money in hand, Chen Ying-bin, the most famous carpenter in the area, during that period, was called in and he designed and supervised a thorough reconstruction of the temple site. This construction began in 1908 and was not completed until 1912.[15] The famous temple that stands today at the center of Peikang which is the site of Taiwan's most famous and well-attended pilgrimage festival was the one so carefully and lovingly restored and expanded during this span of years between 1908 and 1912.

The Tachia Chen-lang Kung. The early evolution of the temple that became Tachia Chen-lang Kung, presents us with two different and overlapping patterns of development. The first is a pattern related to that of the early Ming or Ch'ing settlement and to the expansion of the Minnan presence north into the Changhua area and then further north into the area along the west coast of the island between the present-day port facilities of Taichung and the Hsinchu areas.

The origins of the Tachia temple are obscured by the existence of contradictory accounts of the founding of the temple that are found in local and regional gazetteers from such locales as Tamsui and Hsinchu. One such account states that the temple was founded during the final years of the Yung-cheng era. Another states that the temple was founded in the thirty-fifth year of the Ch'ien-lung Era. A third suggests it was begun in the fiftieth year of the Ch'ien-lung Era. Another set of related questions are these: Was there the one temple we now have and also another Mazu temple and, if that was the case, did these two temples merge to become the present-day temple that dominates the center of town, the T'ien-hou Kung. Questions also exist concerning which temple the Tachia temple is related to as a daughter temple. Some suggest direct links to Meichou while others suggest it was connected to a Mazu temple in nearby Hsinchu. In the 1970s and 1980s, scholars writing in major scholarly journals that deal with Taiwan, specifically *Taiwan Feng-su* and *Taiwan Wen-hsuan*, examined the local gazetteers and presented their views on the problem of origins.[16]

A more recent account of the issue of origins can be found in Chang Hsun's dissertation on the Tachia pilgrimage. Dr. Chang carefully avoids the problem of origins and gives a clear picture of temple development in the context of the town's development over time. This shows where the temple—one not in a seaport—fits into the larger network of temples to the goddess.[17] When the time came for those scholars employed by the temple's administrators to write a narrative history, they decided to take a middle course by laying out the details of the contending accounts and then presenting their own synthesis of what did happen. As I have suggested, one account suggested that the temple was founded in the Ch'ien-lung Era. A second account suggests that the temple was founded in the Ch'ien-lung Era but in a different year and that it was connected to a Hsinchu Mazu temple. The third account suggests that the temple began in the Yung-cheng Era and that it is not a client of any already existing Taiwanese temple but is a daughter temple of the Meichou *tsu-miao*. This is the version that the Chen-lang Kung's official historians are most comfortable with and they start their history here. However, they do not negate the other accounts. The second is related to a pattern of development that moved from the north—the area of the Taipei Basin and the river-mouth port of Tamsui to the south. This later pattern becomes discernable in the late 1700s and becomes even a major factor once rail lines are established and the railroad is pushed south from the Taipei basin.

The temple began, according to the the *Steps of the Pilgrims* account,[18] in the decades after the area was settled in by members of two families from Fukien, the Lins and the Chang and one family—one may assume Hakka, the Chien, from Kuangtung. These families moved into an area on the border between Han-Chinese and the Pinpu tribe. This area was then organized into eight communities. These are Taichiatung, Chipei, Tehua, Shiungliao, Wanli, May Mayou, Fangli, and Tungshiao and became known as Ponshan. Over the course of the next three decades the area developed. A port, Taan, was built and this gave the area a direct connection with the Fukien homeland a hundred miles to the west.

It was to this frontier region that Lin Yuan-yu brought his family. He and his wife and children had lived on Meichou island in Putien county of Fukien and relocated to the Tachia area. As one might expect, a man who lived on the island where the *tsu-miao* was located was a dedicated follower of the goddess. Under the guidance of the goddess, Lin obtained a Mazu *shen-hsiang* from the Ch'au-T'ien Kung in Peikong, and began praying to the goddess day and night. Fifty-three others from the village joined him and a small congregation now took shape. The goddess seemed to bless this small group and in response they called upon a *feng shui* master and on the right day, in the tenth year of the Yungcheng Era (1732) they established a small temple providing the *shen-hsiang* with a permanent home.[19]

To the student of evangelical Christianity, the event just described seems to have been the Mazu cult's version of a religious revival. The evangelical and Pentecostal churches that were formed in the United States in the nineteenth and twentieth centuries were founded in a manner similar to the founding of the temple. God gives his signs and his blessings, prayer becomes a staple in the lives of the followers and church building and institutionalization follows.

This was the beginning. A temple community did indeed evolve during Yuncheng and the Ch'ien-lung Eras. As the community grew, the original tiny temple was inadequate. The members of the community collected funds and a new temple, the T'ien-hou Kung, was constructed in the year 30 of the Ch'ien-lung Era. Later in the century, Chen Fu-chung, a degree holder from the Tachia region of the Tansui Bureau (fu) of Taiwan Prefecture and a local member of the gentry, Lien Kuan-san, gathered funds once again, and had constructed on land a new temple contributed by a man named Chou Hwa-lung. This new temple was called the Chien-lan Kung. The newly rebuilt temple's name means "to conquer the ocean surges." It was able to develop and prosper as the area around it developed.

This temple's subsequent history reflects the growing complexity of Taiwan's development in the nineteenth and twentieth centuries. During the nineteenth century, temple elders began the practice of pilgrimage to the Ch'iau-t'ien Kung. Such pilgrimages became a central element in the life of the temple in the decades that followed.

The development of the railroad system began to have an impact on the area during the early decades of the twentieth century. The mountain railroad was completed in 1908 and the coastal line was finished by 1922. These systems linked north and south. The major roads paralleled the coastal railroad. Tachia profited greatly from the development of the transport infrastructure.[20]

The temple profited as well and as a result was expanded to meet the needs of the community. In 1888, a major repair and expansion took place. In 1935, yet another repair took place. This second repair requires a more detailed discussion.

The temple repair work was necessary because a major earthquake had struck central Taiwan. This earthquake, Chang Hsun shows us, became a watershed in the development of the temple and in the way the temple was perceived in the eyes of people in the Tachia area and beyond. Before the earthquake struck members of the extended Penshan temple community had gone on pilgrimage to Peikong. The earthquake hit the area on the nineteenth day of the third month, and the next day thirty-seven aftershocks struck the area. However, only one person was killed in Tachia and none were killed in Taan and Waipu. However in the nearby towns of Houli, there were 400 fatalities and in Ch'ingsui, too, there were 400. Many people from Tachia, Taan, and Waipu had participated in the Chenlang Kung-organized pilgrimage. However, only a few people from the other two areas, also associated with the Chenlang Kung, had participated. Observers felt that the goddess had protected the towns of the faithful pilgrims but had not protected the towns of those who had not been willing to go on a strenuous and time consuming pilgrimage to the south. In the years that followed, the pilgrimage attracted more and more people from the three towns. However, in Houli people continued to neglect the goddess and her cult.[21]

In the years that followed the temple continued to prosper. Though some space was lost in 1937—removed to make way in the face of a town reconstruction plan. A new open-air area was provided in the front of the temple and this became an open-air market and a meeting place for people of the town. The temple continued to expand during the Retrocession. In 1962, for example, the northern room, lost in 1937, was rebuilt. In 1971, a two tiered drum/bell tower was constructed, new niches for gods were built and "colored paintings, roof clippings etc. were made and the court had a new presentation (space).'[22]

Nine years later, in 1980, the members of the temple's administrative committee made an important and painful decision. They decided that the structural integrity of the temple was weak and that it had to be rebuilt. In the next four years they dismantled the temple, rebuilt the structure and then used the valuable pieces of the old temple in the new, and sounder, structure. This new, yet old, temple is what is seen today.

The modern history of the temple does not end here, however. The main focus of temple life had become pilgrimage, and this was seen throughout the 1980s and the early 1990s. The major pilgrimages to Peikong and later to Hsinkong have been described in the temple's major publications as have the temple committee's pilgrimages to Meichou and Kangli in Fukien. Chang Hsun has studied these pilgrimages in her dissertation and other works. What these sources and analyses demonstrate is that these large-scale pilgrimages serve as a way of connecting the Chenlang Kung's humble past and magnificent and influential present.

Messages to be Found in the Medium of "Time/History"

Now, we must focus upon the "message" component of the "medium/message" dyad. What are the messages that these histories—these reconstructions of each temple's quasi-mythic and actual past—convey to the Mazu faithful?

The first message conveyed in these histories is the relationship each temple has with the most important of the *tsu-miao*, the T'ien-hou Kung on Meichou and/or with one of the other *tsu-miao*. A major temple must show it is related to one or more of the fonts of spiritual power that are found in Fukien. The historical or mythohistorical record of such contact is meant to show the public that that temple began with a sharing of Mazu's *ling* as transmitted through the *shen-hsiang* of Mazu that the temple received. These accounts also usually show that the temple received the *shen-huo*—the fire or incense that the missionaries from Meichou (we may call these temple founders the missionaries of the Min-chien—the popular-tradition) brought with them. I would argue that establishing evidence of a link with the Putien (or the Ch'uanchou) *tsu-miao* is part of a process of legitimatization the temple committees feel their temple has to go through. The temple's history, as recorded in one or more of its publications, is the medium that conveys this message of links with the spiritual center of the cult and thus with the goddess herself.

This linkage with the *tsu-miao* is important in another way: by demonstrating that the temple is linked with Meichou, the temple shows it is a link in a chain and it, in turn, has the power to create its own network of daughter or client temples. And thus the temple, like the mother temple, can claim to be a center of pilgrimage and festival. By showing the linkage it also shows that it was one of those special temples blessed by the goddess and is thus a place of special power. The major temples also try to demonstrate that Mazu has continued to bless them. These spiritual phenomena—the role of the goddess in the origin of a temple and the continuity of gifts from her during the life of the temple—suggests that there are a second and a third message to be found in the temple histories.

This second message, conveyed, like the first, in each temple history is that the goddess started the temple with an act of personal intervention. The pattern that one finds is this: the goddess tells a devout follower that she, in the form of her *shen-hsiang*, wants to be in a certain seacoast town. The *shen-hsiang* is then brought on land and a home is built around her/it. This is the pattern we see in the accounts of each of the three temples we have examined, though with slight variations. The goddess wants her will obeyed. She wants this not for herself, however, but for the people of the community who will receive the blessings given to those who worship her and who feel her power. This moment of founding of these temples is an important one. Even where the historical record is unclear, the link with the mother temple and the command that a *shen-hsiang* be put in place as the focal point of the goddess's *ling* is presented in that temple's history. Thus I prefer to see these accounts and these descriptions of events in terms of a mythic history, regardless of the efforts of the temple's chroniclers to root them firmly in the actual past.

The power of the goddess continues to affect a temple in the years, the decades, and the centuries after its founding, and continues to play a role in the development of that temple and its community. This is the third message we find. The Peikong temple, for example, has its famous nail and the story of the goddess's power that is attached to it. The rebuilding of the Hsinkong temple is linked to the impact that the goddess had upon the life of a major Taiwanese military figure who responded to the goddess's blessing in a powerful way. The Tachia temple has its own stories of the power of the goddess. These, however, are not linked directly to the temple, but are tied to the pilgrimages that people from the temple made to Peikong. What this shows—and this is something that we will explore at a later point in this chapter, is the role of pilgrimage and networking and the part that the goddess is said to play in such rituals of faith.

There is a fourth message that the histories convey—that each temple has been an evolving and expanding religious institution that continues to serve the goddess. And the very record of progress that is suggested in the temple history is dramatic proof that Mazu continues the process of blessing her followers, the temple's home community, and the communities of the daughter temples that are part of the temple's network. The historical record links the past with the present in a powerful way, and, in turn, is meant to convince each reader/believer that he or she can feel part of this specific chain of being by believing in Mazu and by taking part in its religious life or in the religious life of one of the client temples.

The "history" medium/message dyad does not stand alone. I see it as one part of a larger set of dyads, each of which are inter-related and each of which contributes to the larger message of faith and loyalty to a temple that temple administrators want to convey. We will now examine the temple site itself and see it as as medium and source or producer of messages.

Space as the Second of the Medium/Message Dyads

The term "space" is use as a shorthand for the second of the medium/message dyads we will examine. "Space" in this case means "holy space"—the Mazu temple in Taiwan. I would argue that we can consider the temple as a medium. If we study the way the temple is designed and constructed, if we examine how each part of a temple is decorated—for example, what icons and motifs are used—what the *shen-hsiang* of the goddess, her guardians and related gods and goddesses such as Kuan-yin are like, as well as the way the holy space is arranged and utilized, we can find sets of messages that the temple committee uses in its spiritual battles with rival temples.[23] While each temple shares certain basic characteristics, there are differences in the structuring and use of space as well as other differences that are products of each temple's history and pattern of development. In this section I present descriptions of the temples we are studying. Only with these descriptions in hand do I turn to the question of messages generated by the temple as "medium."

The Temple as Medium

Each of these three temples—the Hsinkong Fon-t'ien Kung, the Peikong Ch'ao-t'ien Kung, and the Tachia Chen-lang Kung—is an impressive building in its own right. Each temple also shares certain basic characteristics with its rivals and with other Mazu temples on the island. Further, each of these temples also share certain characteristics with other temples devoted to popular gods. The design of most Taiwanese temples is based on patterns defined over the centuries in Fukien.[24] The typical temple found on the island is built along the lines of this Minnan mode. The typical basic temple is a relatively simple structure in its overall design, if not in its actual details. Certain parts of the temple are required by custom and what might be termed the canons of temple construction. Basically, the temple is a functional building designed for worship of a major god/goddess and for worship of those deities related to the central deity.

This basic (or core) temple has roofed-over entrance gates. These gates are set back under the eves of a roof with a reverse arch and are flanked by elaborate columns. Each of the columns has an architectural and a symbolic purpose, as well, and each is given its own name as are each of the segments of the two main gates to the temple. Iconographic figures on the roof also have a religious role and significance. Once one is inside the gates, one sees two side halls flanking a large bricked-over space that is open to the sky. At the center of this space is a raised platform where the major deity and her or his guardians are found. Here, one also finds a carved figure, a dragon. One ascends the platform by taking steps located in the center of each of the side halls. The holy of holies is covered with a roof having a traditional reverse curve that matches the roof of the entrance hall. There are a number of roof styles that temple builders can choose from. Temples on Taiwan demonstrate the different styles. In this design, the central sections of the roof are supported by four sets of stone columns. The first set is elaborately carved in a manner reminiscent of the way the columns in the front of the entrance hall are carved. Each of these sets of columns has a variety of purposes. The two sets at the front of the central platform provide an entrance way to the holiest space within the structure. The first of two sets to the rear, flank the table that lies directly in front of the *shen-hsiang* itself. Here the sacrifices to the goddess are placed. The final two columns, set a few feet from the rear wall of the temple flank the niche where *shen-hsiang* is found.[25]

One further comment is necessary: as is true of members of any field or specialty, those who plan and construct temples have developed their own vocabulary for each of the separate pieces that fit together to form the popular (or text-centered) temple. There are differences between the Chinese terms and English words used to translate them. The translators who prepared the Chinese/English version of the diagram of a typical temple that is found in the three Sinorama volumes on Taiwanese temples—and in the larger set of volumes on Taiwanese temples published by the Taiwanese Joint Religious Board, for example used simple and straightforward English words to translate more arcane and descriptive Chinese phrases. The essence of the terms are retained, but their unique flavor are lost. The Chinese characters themselves go far to suggest the nature of this unique temple-architecture vocabulary.

This, then, is the typical, rather basic, Taiwanese temple. A visit to any of the major Mazu temples on the island will demonstrate just how far these temples have evolved from this simple, functional structure. However, a journey to a relatively unchanged or moderately changed suburb such as Anp'ing and to a classic town such as Lukong will not only take one to some of the large, elaborate, and famous temples, but will also allow him (or her) to see temples to various gods, such as Wang Yeh or Pao-sheng Ta-ti, that fit the simple classical form I have described—with the help of the editors of the *Sinorama* compendium.

This basic design provides what I see as the core or central structure of the three major temples we are examining. Each of the temples being profiled evolved out of this core and its administrators added to it, but it is in the temple's central sanctuary that the core temple can still be found, if one searches for it. One must also add, however, that the major temples have been built and rebuilt a number of times over the course of their histories. As each reconstruction has taken place, more elaboration or variation on the basic theme has been introduced by the architects and builders.

The Hsinkong Fon-t'ien Kung

The great temple in Hsinkong, the Fon-t'ien Kung, dominates the center of the old town. One enters the temple through the main gates. Guarding these gates is a pair of stone lions carved with what has been described as "a natural, plain expresssion." The visitor or the pilgrim also sees stone Peony Pillars in front of this gate. These are unique for on them are carved hundreds of phoenix. When one looks above the pillars one sees that between the roof beams are square pecked-shaped boxes aligned in rows. The lower parts of these boxes are decorated with winged four-clawed dragons.[26]

Stepping through one of the temple's main gates one can see the other buildings (or courts, to use the language of the encyclopedia) that make up the temple. One first walks into the bright, open bricked-over courtyard—the light well—and looking toward the walls of the side halls one sees two series of connected three-dimensional picture—human images all involved in activity of one sort or another and all done in brightly colored ceramic.

Passing through the courtyard/light well one comes to the temple's "main palace." It is here that one finds the main Mazu *shen-hsiang*. On the outside of the central shrine, on the walls, are carvings of foliage, flowers, and birds brought together to form complex decorative patterns. Above this hangs a tablet dedicated to Mazu and written by the Yungcheng Emperor. Within the "holy space" are two support beams darkened by incense smoke. Under these beams is an eight-angled glazed light that shines day and night, though it too has a surface obscured by centuries of incense smoke. In front of the *shen-hsiang* is the offering table and on it is placed what is described as a "file-made censer" that was made in year 24 of the Taokuan Era. On the censer is a fine engraving of a dragon. Mazu herself dominates the central position in the niche and she is flanked by the two generals who serve as her guardians, Chien-li-yen and Shun-fon-ehr. Forty-eight holy sticks (spears with elaborately carved heads), each representing a general and used for escorting Mazu when she tours her lands, flank the statues. A final, but still important item one finds in this area, is a tablet inscribed by Wang Teh-lu, the military man who financed the reconstruction of the temple. It reads, "The sages are gracious and Mazu is virtuous."[27]

One can reach the rear "palace" by walking through either of the galleries that flank the central shrine. In this rear structure are *shen-hsiang* of Kuan-yin, Buddha, and A-mi-to-fo. Above the Buddhas and the Bodhisatva is a tablet that was prepared and given to the temple by the head of the Penkong district. On the table that sits in front of the *shen-hsiang* are three stone censers, each carved in a different dynasty. Religion and history merge in this area, as they did in the central shrine of the goddess. In the halls that flank this rear building are shrines to Wen-chang, the god of literature and learning and to Kuan-kung, the god of war, government, and justice.

Other halls or palaces have been built as well. One of these is dedicated to Mazu's parents and another to Wang Teh-lu, the temple's major patron. Administrative offices have also been constructed within the building complex.[28] This is the temple itself. But the great temples needed more space to cater to the needs of the faithful. The administrators of this temple, used the contributions from the faifthful to construct a multi-story building in a eclectic style to house several thousands of pilgrims.

The Fon-t'ien Kung is a structure that has put the town of Hsinkong on the map. A central theme that one finds when exploring this temple and studying it in all its rich detail, is that of linkages with the Taiwanese past. The temple is also a storehouse of iconographic works—stelle, plaques, wall carvings, which have their own special historical significance. One can find the core temple if one looks carefully, and will experience the powerful sense of continuity that links the present to the Taiwan of the Ch'ing and Ming/Cheng Eras, the eras of the island's settlement by Minnan pioneers.

The Peikong Chau-t'ien Kung

The Ch'au-t'ien Kung is an imposing structure—a grand larger than life, turn-of-the-century expression of the Minnan/southern temple-building tradition. It is a building complex that assaults the senses with its vaulting spaces, its color, and its sculptures and its wall decoration with their wealth of detail. One must visit the temple when it is relatively quiet and slowly explore its corridors, "palaces," and gardens, its central shrine to begin to appreciate all that that it means to the Taiwanese.[29]

The temple lies in the center of Peikong, dominating the small city by its occupation of space—it covers a thousand pings of land—and by capturing the attention of visitors and pilgrims. It has also become an essential part of the town and the region's economy.

Some sense of the temple and its spaces can be gained by examining the diagrams the temple administrators include in the *Peikong Ch'au-t'ien Kung Chih (the Peikong Ch'au-t'ien Temple Gazeteer)*. One can see that the temple's structure is vast. It is surrounded by an oval outer wall and is divided into a number of separate sections—halls or "palaces," and which are further subdivided. The vast majority of space is what we might call worship space—space where the faithful can pray to their patron god or goddess and offer incense to them. There is also a room where the pilgrims can wash and prepare themselves for meeting either T'ien-ho or the other major and/or minor gods they have come to see.[30]

Let us now tour the temple and see its many splendors through the eye of the pilgrim or visitor. One can easily imagine the awe of the pilgrim as he gets his first glimpse of the great temple and as he moves with his group slowly up Chungshang Road to the outside courtyard and the three great gates. There are other ways to enter the building—there are four entrances to the temple—but this is the one that most visitors and pilgrims would use. The temple has three gates, the Lung Men (the Dragon Gate) to the right, the Hu Men (the Tiger Gate) to the left and the Shang Chuan (the Mountain/Rivers Gate) in the center.

It is the central gate that the pilgrim/tourist would most probably walk through. Before he did so, however, he would take a close look at two large pillars that flank the gate. These pillars are covered with carvings of great dragons that spiral their way upward from the pillar's base. Once through the entrance, the pilgrim would see four additional support columns and walls covered with richly detailed carvings of birds, foliage, and flowers. The temple's public spaces are filled with example after example of such sculptures as well as sculptures of people involved in activity of all sorts. It is this rich array of popular art that gives the temple much of its appeal and its status as a "2nd grade relic."

Beyond these sculpture-covered walls and panels he would see the open courtyard that stands in front of the central palace. Here, he would find an incense burner in which he would light the sticks of incense and pray to Mazu as she sits before him in the holy of holies. The main *shen-hsiang* dominates the large god niche. The ancient and powerful *shen-hsiang* the pilgrim would pray to and bow before is not alone, however, for near her are twenty-nine other statues of the goddess. Near her also stand the gods we most often see in a Mazu temple. We would see for example, her ladies in waiting and the two generals, the all-seeing general, Ch'ien-li-yen, and the all-hearing general, Shun Fon-erh, standing near the goddess and protecting her from harm.[31] On both sides of the "main palace" are important relics such as a pair of clocks and a drum. Wang Te-lu, the patron of temples in the Peikong and Hsinkong areas is the historical personality associated with these relics.

If the pilgrim, or the visitor, moves beyond the central hall he (or she) would come to three open spaces which serve as courtyards for three small courts or temples. The central temple in this group of three belongs to the Buddha. The temple on the left is the home of Wen Chang, the god of the scholars. The last of these, the temple on the right, is for the San Chieh. There are niches for the *shen-hsiang* and these are flanked by decorated columns. Moving further toward the rear of the temple complex one would come to a large courtyard. Yet another group of "courts" are arrayed along the side and back of this open space, the largest that one finds in the Ch'au-t'ien Kung. In the place honor is the court/temple devoted to Mazu's parents. Such temples to the goddess's parents are often found in the great Mazu temples and this is a second example of the practice. Often, there are other *shen fang* (god halls), but the diagram of the temple does not tell us what god has his or her *shen-hsiang* in these halls.

As the pilgrim strolled through these halls, courtyards, and courts, he would see a wealth of ancient tablets and plaques, he would find equipment that was used by Shu-pi, the monk who helped found the temple, a stone dragon pillar presented by the Ch'ienlung Emperor, stone censers from various periods, and a host of other similar artifacts.[32] That each of these was presented to the administrators suggests the power of the temple and its significance in the life of Taiwan.

The pilgrim, or, more likely, the interested tourist might go from these rear halls to the outer halls and/or open spaces that one finds between the temple's circular outer wall and the inner wall that is part of the roofed structure of the temple complex. On the left side one finds the Golden Pagoda. This is where the paper money to the gods and ancestors is burned. On the right side are a number of connected smaller halls. The first of these is one of the places where the pilgrims can wash and prepare themselves. Next comes the Gong and Drum Tower and then the Pan Kung Ting (Moving Lord Hall). Here, the palanquins that carry the goddess around the city of Peikong and elsewhere are found.

This completes our tour of this old and famous temple. The temple is a place where faith in the goddess and in the other gods and goddesses of the Minnan pantheon is palpable and alive. If I may wander from the path of rationality and objectivity here, I would suggest that, even as an outsider, I could feel the power of the place and of the gods who occupy it. During my visit to the temple in that spring day in 1980, I witnessed and photographed a *chi-kong* (shamen) in a trance, in the courtyard in front of the main palace. I did not know then that I would study the popular tradition, for I was just beginning my study of Christianty in Taiwan. What I did know was that I was witnessing something I had never seen before and that I was in the midst of a throng that knew what they were looking at and were feeling its power during this moment of the meeting of different sources or instruments of *ling*.

In a subsequent visit, in the summer of 1992, now that I was seriously involved in an attempt to understand the Mazu cult and its role in Minnan life, I toured a quieter temple. Though the mass of humanity was gone, pilgrims and worshippers were still on the scene as that summer day wound down. Again I felt the presence of power. It was the power of the site and of the goddess who made that site holy. But there was something else there—it was the power of history that had taken place there, and it was what I can only describe as the psychic residue of a place where millions over the centuries have come to receive the blessings of the Holy Mother.

To study religion one must consider religion as a non-rational activity seriously and one must, at times, open up one self up to that non-rationality. It is at sites such as the Ch'au-t'ien Kung that one can open one's self up in this manner and then, only later, attempt to verbalize—while trying not to rationalize or really explain—what one has experienced.

The Tachia Chen-lang Kung

The Tachia Chen-lang Kung lies well north of the core area of Chiayi. We have seen that the temple's social and geographical environments differed markedly from those of the Hsinkong and Peikong temples. Because of these differences the Tachia temple was forced to mold a very different role for itself in the life of its local community and in the life of the Taiwanese Mazu temple community. Its development as a socio-religious institution reflected this different pattern. Pilgrimage and service to the community are the key factors in the development of this temple. These two factors are reflected in the temple's use of space and the way the temple's structure evolved and was transformed over the centuries.

The temple is in the business center of Tachia, the economic heart of the larger Penshan area. It lies along a major north–south thoroughfare. This location is fitting for it reflects the fact that the temple's evolution was part of the process of Taiwanese movement and development from both south to north and north to south.

The temple stands near the intersection of Shuen-t'ien and Chian-Kung Roads. One looks to the east and would see two banyan trees that stand on both sides of the temple. The outside walls are of Kwan-yin stone. Sculptures of the twenty-four filial sons are on this wall. The entrance gate of the temple is guarded by two grand stone lions which have stood there for 250 years.

The temple itself is built in the three-stream/five entrance style. Standing back from the main entrance one will see the colorful roof of the temple. It is filled with figures such as birds and animals and there are flowers as well, all executed in ceramic. At the entrance gate one would see stone pillars with dragons spiraling up them, each of these fabled beasts are carved with great artistry. In front of the mid-gate are cow stones.

When the visitor/worshipper passes through and looks back he would find the Door God staring at him. In the open space before the central palace one would see a dragon well with elaborately carved nets. Beyond the celestial well one would see the stone pillars, elaborately and ornately carved columns that support the central holy space of the temple. In the celestial well, there is a mammoth-sized censer holding burning incense sticks and the ash of the incense already burned in honor of the goddess.[33]

One can ascend stairs from the celestial center to reach the central palace, the Chen-lang Kung's holy of holies. Here, one would be greeted by a set of stone lions surrounded by a jade rail that has been in the temple since late in the Ch'ien-lung Era. Once the visitor is in the central palace itself, he (or she) will usually take in its wonders, before he begins to pray and ask the goddess for her blessings. This splendor—this expression of magnificence is an important message conveyed by the temple-as-medium. Above the worshipper is an elaborately carved roof that is filled with images of fairies and other adornments. On the walls of the central sanctuary are sculptures. One depicts Mazu ascending to heaven, while the other depicts her guarding her country. The niche where the *shen-hsiang* sits is elaborate and ornate as well. Above the niche is a plaque with the words of the Ch'ien-lung Emperor on it. There are various Mazu *shen-hsiang* in this central space. Some are from Meichou while others are from Kangli. Each, according to the authors of *In the Steps of the Pilgrim*, have kind countenances. These *shen-hsiang* are surrounded by the guardian generals and by the maidservants of the goddess. There are other niches in the central palace. In one is the goddess who protects pregnant women. She is to the left. To the right is the Chialan Buddha, one of the many who is seen as a protector of people in the Tachia area.

When one moves out of the central palace and steps left one would come to the palace devoted to Kuan-yin. Here, the *shen-hsiang* of both the goddess of mercy and Wei-tou, the Chaste Mother Buddha, are found. There are other *shen-hsiang* here as well. On both sides of the hall there are Lohan. There is a further subdivision of this space into a right and left hall.

The left hall is further subdivided and here gods of the constellation are worshiped. There is another hall to the right of the central hall. Here, ones would find the *shen-hsiang* of Kuan-kung and the god of agriculture. The roof of this hall or palace is particularly ornate with dragons and phoenix in abundance. The last structures a pilgrim or visitor would find within the temple complex are a bell tower on the left and a drum tower on the right.

Although the Chen-lang Kung is a new structure, it is still constructed in the southern/Minnan style; much of the older temple has also been preserved. Many of the relics have been preserved and are displayed prominently. These pillars, censers, and plaques tie the old temple to the new.[34]

The Message of "Temple Space"

Each temple's definition and use of its holy space is meant to serve as a medium and to send a variety of messages to the believers in the goddess. But what is the general nature of such messages? We must begin by recognizing that one can find important commonalities in the way that each temple has defined the nature of and has utilized its holy space. The general layout of the temple is similar. The central palace too is set up in a similar way as are the other halls or palaces dedicated to the related gods and the actual choice of the gods or goddesses. Finally, the worship spaces, where individuals and pilgrimage groups can praise the goddess, commune with her, and receive her blessings, is similar. What such similarities mean is that at least some of the messages conveyed by the temple-as-medium will also be similar.

A first message common to all the temples is this: each of these temple's continued growth and prosperity can be seen as evidence of the living presence of Mazu. Her presence can be seen in the glory of the temple that has evolved over the centuries. It can be seen in the detail of the temple's carvings, and in the relics of Taiwan's and China's past that are displayed in the various holy spaces.

This presence is most clearly seen in the temple's central palace. Here, in the "holy of holies" one can find the great *shen-hsiang* of the goddess herself. Here we have the guardian generals who protect Mazu. Here, we have the ladies in waiting. Here, we also have many of the other *shen-hsiang* of the goddess that the temple possesses. Some of these were presented to the temple. Others were obtained when the temple's representatives went on pilgrimage to the *tsu-miao* in Putien County and/or Ch'uanchou County—Meichou, Kangli or to the city of Ch'uanchou.

The living presence of Mazu is most clearly felt during the ceremonies that culminate a pilgrimage from a daughter temple to a mother temple. The daughter temple's various *shen-hsiang* are passed through the incense smoke and fire and then are placed near the main *shen-hsiang*. It is now that the *ling* of the goddess is at its strongest and it is felt by those present. This is temple as medium and temple-centered rite as message.

A second common message conveyed by temple-as-medium is that of identity. The nature of that message is not simple for the type of identity that is conveyed is a multi-leveled one. One such level is the local/regional one. The individual who is on a pilgrimage to the mother temple experiences both the power of the goddess as he (or she) bathes in her *ling* and also experiences a sense of comradeship with those he is with. There is a shared sense of identity with those one is with, in that spiritually charged environment. And there is a moment of bonding and sharing that goes on—bonding with one's fellow pilgrims and bonding with those who administer the temple and have helped to create or maintain this sanctuary where one can worship the goddess and receive her blessings.

But there are other levels of identity that the Mazu is part of, and these second and third higher levels also convey the message presented by temple as medium. There is, first, a sense of a greater Taiwanese identity. True the goddess is linked to a given temple but she is also a goddess worshiped throughout the island. In praising her and in accepting her blessings one is sharing in a trans-island act that reinforces one's sense of being Taiwanese.

Mazu is a goddess worshiped in Fukien—she is a Minnan-wide deity. Many Taiwanese have undertaken pilgrimage to Meichou and some have traveled to Kangli, and even more to the T'ien-ho temple in Ch'uanchou. At these moments, in Meichou and in the heart of the *tsu-miao* complex, these pilgrims share an experience of the religious devotion and an experience of heightened ethnic awareness. And the individuals on pilgrimage are well aware of the multiple levels of their own experience, and are willing to share their insights and their joy in being able to participate in a ritual on Minnan/Fukien holy ground.

The temples in Taiwan have done much to promote this idea of the expansive power of the goddess. They have reconnected with the Meichou temple and some temples such as Lukong have actively promoted tourism/pilgrimage and developed package tours to Meichou and related religious sites in Minnan Fukien. The three temples we have studied have also begun to play a part in the life of the Meichou *tsu-miao*, contributing funds and helping officials of the site to reconstruct the area. This has been taken a step further as temples in Taiwan have supported the construction of new temples on the Meichou site. Finally, the new and crowning glory of the island is a gigantic statue of the goddess that has been constructed with funds from Taiwan. Meichou has, thus, become a center of a greater Minnan form of religiosity and a new arena for temple rivalry and temple conflict.

But one need not go on a pilgrimage to become aware of the larger dimensions of the cult and its influence. The major temples discuss pilgrimage and the greater Minnan role of Mazu. One can read in these sections of temple gazetteers and more secular descriptive works, an attempt to articulate a spiritual form of cross-the-Taiwan-Strait identity that links goddess and reason. The worshiper reading these texts will understand the message and will, the authors hope, also understand the role of the goddess in promoting a larger vision of pan-Minnan identity. The pilgrimage then became the pivotal event in the inscribing of this sense of spiritual/cultural identity. Let me end on this cautionary note: I am well aware of the suggestive quality of these ideas of linkage and the interaction between religious and national identities. The fieldwork I have done, and those interviews I have conducted, supports their validity, but more remains to be accomplished and more hard data remains to be gathered.

The identity-related message takes yet another form in the "temple-as-holy-space" medium and conveys a broader vision of identity and religious reality. Other gods and goddess are worshiped as well and these are deities that also speak to the nature of both Taiwanese and Minnan identities. One must add that they also speak to question of Han identity. They are powerful in their own right and their presence increases the power of the temple. But in worshiping them the individuals are also linking themselves to a greater vision of their temple and Taiwan's past. These are not just local gods, nor are they merely Fukienese gods. They are Chinese gods. Thus, a Chinese universe opens up to the worshiper. The temple stands as microcosm of this expanded universe and those who pray here enter this universe, and partake of its wonder and its power.

Word: Temple Publications as Medium

"Word" is the shorthand I use for the third and final medium/message dyad. Word, here, is meant to suggest a temple's publications. Those temples with large Taiwan-wide followings usually prepare and print different types of publications. These include both small-format pamphlets of different kinds, and large-format hard- or soft-cover books. Both the Peikong and the Tachia temples provide the religious public with books that are prime examples of both the large- and the small-format. Each temple produces this variety of works as means of engaging the enemy in the battle for the public's attention. Such books and pamphlets become both the medium through which a variety of messages are transmitted and the messages themselves, by the very nature of their appearance and by the nature of the care and expertise that went into their production.

The materials published by the rival temples in Hsin-kong, Peikong, and Tachia fall into either one or both of two distinct categories. The first of these categories contains pamphlets and small-format books that number ten to thirty pages in length. Such works are usually 7.5 by 5 in. or 8 by 6 in. pamphlets that contain anywhere from twenty to thirty-seven pages or more.[35]

The Hsin-kong temple's pamphlet and the Peikong temple's small-format pamphlet are both typical of this type. The Peikong temple's pamphlet contains forty pages of text and four pages of photos. It begins with a general introduction of why the pamphlet was published. Its first substantive section is a biography of Mazu and a chronological record of the development of the Mazu cult on the Chinese Mainland. This is followed by a history of the temple in Anp'ing, a history that first places the development of the temple within the larger history of the area during the era of Chen Che'ng-kung, and then traces the development and growth of the temple complex itself. This is followed by a narrative of the temple's development during the Ch'ing Era. The final section traces the evolution of the temple—and the area—from the Japanese period to the time the pamphlet was edited and published in 1975. The material is carefully presented so as to make the case for the temple's historical role in the development of the cult, and its present importance as a thriving institution that remains a stronghold of the Mazu faith in a major early center of the faith.[36] I see the small-format of this type as a variety—on a much smaller scale, to be sure—of the type of gazetteer that one makes use of when studying local history. I do not want to take this too far, but I think one can talk of these documents as temple gazetteers.

The second category of temple publications consists of far more elaborate, large-format books with upward of a hundred pages of text and colored illustrations and photographs. These expensively produced volumes are designed to present the reader with word pictures and actual images of a temple's history, its contemporary ritual life, and the nature of the temple's present-day leadership. Pilgrimage to Fukien and linkages to the Put'ien/Meichou *tzu-miao* are also covered in these materials that the temples themselves publish and distribute. The great temples have the financial resources to produce and publish such large-format books. The Peikong Ch'ao-t'ien Kung and the Tachia Chen-lan Kung are two of the most important Mazu temples on the island that now stand as competitors. Such books provide the members of each temple with a rich store of information about their goddess and further, provide the believer with evidence of the power of the temple they support with their labor and with their hard-earned money.

The Messages to be Found in Publications

As I have demonstrated, these temples with Taiwan-wide followings publish different types of texts. These include both small-format pamphlets, of different kinds, and large-format hard- or soft-cover books. Both the Peikong and the Tachia temples, for example, provide the religious public with books that are prime examples of both the large- and the small-formats. These published works serve as a means of engaging the enemy in the ongoing battle for the public's attention and the escalating struggle for influence.

But what of the messages produced by this medium. That part of the dyad will now be examined. The large-format works published by the Peikong and the Tachia temples produce what may be seen as comprehensive sets of messages. These publications are elaborately produced large-scale works and by their very existence—by their size, their heft, their elaborate production with the many colored photographs and varied styles of characters—make a statement, or in the terms we have used in this chapter, presented a strong message. The message is: these temples are serious players within the Mazu community and they want the believers and the general public to know this. They have influence, they have power, and they have been blessed by the *ling* of the goddess herself. This is demonstrated by their ability to produce such a magnificent volume—a volume that defines who and what they are.

But the presentation of the texts go only so far. A second message is contained in the contents of the book. This, too, is a message of power and influence. However, now it is the substance that makes the point. The major temples use these books to sum up what they were and what they are. The objective, here, is to present themselves in most positive way to the Taiwanese public. They call upon known scholars to write these works and then give them the material they need to complete this task. The scholars are backed up by photographers who are able to capture the temple and its activities in the best possible light. When they produce the books and publish them, they spare no expense. Thus, these books, like the temple's history and its holy space, serve as a way of bringing the temple to the public eye and winning over the Mazu faithful.

I would suggest that a volume such as that produced by the Ch'au-t'ien Kung or the Chen-lang Kung is perhaps the most useful form of medium, for such a book makes each of the mediums accessible to the public and presents the history-as-medium in an accessible form using photos to bring that history to life. Furthermore, the detailed diagrams, photos, and text about the temple that is contained in such a book also serve to make that temple a reality in the mind of the believer or the lay person.

Such books do two things. First, they serve as a glorious form of advertisement—making the case for the temple's unique past and arguing that the visitor or the pilgrim should visit the temple to be able to take advantage of all that it has to offer as a religious experience and as an experience of regional, Taiwanese, and Minnan identities. One can visit a temple such as the Ch'au-t'ien Kung or the Chen-lan Kung and, with the knowledge gained by reading these books, know what one is looking at and have a deep sense of why the temple is as important and glorious. And when the pilgrimage or the visit is over, the traveler can pick up the book once again and use it to recall what experiences he or she had at the holy site.

The Fon-t'ien Kung also publishes a book, or to be more precise, a small-format pamphlet. This book publication is much smaller in scale than the ones produced by its rivals and its production values are of a lower order, thus making the temple less competitive in the media battle that takes place among the major Taiwanese temples. However, the temple's booklet does convey a good sense of the history of the Fon-t'ien Kung, thus making a strong case for the temple's precedence and its place in the hierarchy of Taiwan's Mazu temples. It also conveys a word picture of the temple that is written well enough to convince the pilgrim or the visitor to go to this old and venerated holy site and pay homage to the goddess.

These books are meant to attract visitors and pilgrims. This is their major purpose and that is the message they convey. Their purpose is also to make the case for these temple as major forces within Taiwan's Mazu community.

These books have a third purpose and a third message to present: it is to argue the case that to worship the goddess Mazu is to perform an act that reinforces one's sense of being Taiwanese. These books link a temple to the history of Taiwan and Fukien. One reads these books and becomes aware of the fact that as one celebrates at these temples one is becoming a link in a long chain of spiritual being. And if one takes the next step and goes on a pilgrimage to one of these temples, one realizes in one's heart that such a linkage to Taiwan and to its glorious past is real and palpable.

To read these books is also to become aware that one is a member of a larger Minnan spiritual/cultural universe. These books link their temples to the ongoing process of reconnection with the ancestral temples in the Fukien homeland. The temples have pioneered this renewal of relationships, and the books contain evidence of this effort. Furthermore reading these books and seeing images of pilgrimages to Mazu might spur believers to go on such pilgrimages themselves. Thus, the books make the case for a larger sense of identity and a larger sense of personal identification with the China that lies on the other side of the Taiwan Strait.

Thus far, I have described temple messages that are common to the three temples. Yet, we have a problem here. If we find temples producing common messages, messages they all share, how are such medium/message dyads used when temples compete with each other for the loyalty of other temples and the loyalty of the populations of the Mazu faithful?

My answer is this: while there are common messages or sets of messages that each of these temples present, there are also very special messages that each temple presents in ways that demonstrate that it is a unique entity. Such messages of distinctiveness are embedded in the seemly similar messages that are produced. This means that each temple will produce messages that allow the public to differentiate it from its rivals, and that these messages are to be seen as both text and subtext when a temple lays out its history, defines the use of its holy space, and presents its message in a published form.

The differences between these temples—their histories, their use of holy space, and the verbal presentations of themselves—are indeed profound and they emerge in the messages even as the common themes are laid out. As a way of concluding this section, let me suggest how the temples we have focused upon differentiate themselves from their rivals even as they present messages that seem similar to those of these same rivals.

The Hsinkong temple sees itself as the oldest link to the Taiwanese past and to the Taiwanese people's relationship to their ancestors in Fukien. This theme is one that is carried through and is visible in each of the three mediums I discussed. And to the temple's good fortune, scholars and reporters have picked up on this theme and have devoted articles to the issue, making a strong case for the temple in Hsinkong. The Fon-t'ien Kung is a glorious building. The temple as an institution is one that has contributed much to Hsinkong's quality of life and the town's recognition as both an important religious and historical site. These special qualities of the temple are presented in the three mediums the temple committee employs, as they bring their messages to the general public.

The Peikong Ch'au-t'ien Kung's administrators see their temple as an example of the great power of Mazu. They perceive its growth and its prestige as evidence of the working of the goddess in Taiwan—the way the goddess's blessings are manifested. The temple's administrators also believe that their temple is closely tied to the *tsu-miao* in Meichou. These are the major themes that temple administrators stress as they create messages for the three different mediums of time, space, and word.

The temple administrators use the historical record of the temple, its holy space and its well-produced and well-written publications, to demonstrate these themes—the connections to the *tsu-miao*, to the temple's own glorious and sometimes miraculous past, and its power and prestige in the modern day. They also use these mediums to demonstrate the goddess's power and her influence on the people who worship and serve at the temple. They have been able to convince people of Taiwan who worship at client temples that their mother temple, the Ch'au-t'ien Kung, is a key link in the chain that connects their temple both with Taiwan's past and with the *tsu-miao* of the cult in present-day Fukien.

The themes stressed in the messages presented by the administrators of the Chen-lang Kung in Tachia are different from those of the other two temples. They are intended to inform the worshiper and temple member that their temple's past has been different from the other major temples. They demonstrate in their record of the past, in the use and reuse of temple space, and in the books they produce that their temple's role in Taiwan differs markedly from roles played by other mother temples and rivals.

Administrators of the temple discovered in the nineteenth century that for some temples which were established well after the pioneering decades of the late seventeenth century, the act of pilgrimage could be transformed into a central and consuming reality in the life of their temple community, one that gives that temple great visibility and recognition. The tactic employed in the official history of the Chen-lang Kung is a recognition of this reality. While the temple has become rather spectacular, it is the yearly act of pilgrimage that gives it the prestige its administrators want it to receive.

To sum up, one can think of the medium/message sets that the three temples use as implements of persuasion. The medium conveys the message and it is that message or set of messages that are intended to sway the religious public and convince them of the superior power of a particular temple's ability to encapsulate and manifest the greater glory of Mazu.

Patterns of Rivalry and the Process of Identity Formation

The rivalry between temples, and the conflicts that are a product of this rivalry, take place in the socioreligious realm: the battlefields are the temples and the temple communities that dot the Taiwanese urban and rural landscapes. These temples are places of religious/cultural struggle, as well as places where one communes with the major deity or the related deities whose images—*shen-hsiang*—are found there. These temples are grand places. Some are relatively new structures—they are rebuilt versions of temples that have long stood on that same spot. Others are stately older structures whose central worship spaces are blackened by the incense that has burned over for centuries. In either case, these temples have long histories and have connections with one or another of the major Fukien *tsu-miao*.

The strategists and tacticians—and the combatants—in the struggles between the temples are the members of the committees that administer these temples: they are individuals whose lives, and, in many cases, the lives of family members before them, have centered on these temples. Such committees represent their temples in national and international gatherings and on pilgrimages. They run a temple's operations—such operations that can be quite extensive and multi-faceted. They control the sometimes enormous funds of their temple. They plan the various festivals and ceremonies that attract hundreds, thousands, tens of thousands, and, sometimes, hundreds of thousands of believers. Anyone who has been to Peikong during the Mazu birthday festival will attest to the fact that a sea of faces confronts the onlooker. They arrange pilgrimages to the mother temples—usually the Meichou *tzu-miao* as well as the tours that are often part of these pilgrimages.[37] They also prepare and publish various documents ranging from two- or three-page brochures to large-scale cocktail table-style books. Finally, they plan the various tactics and strategies in the struggle for believers.[38]

But what is nature of the rivalry and the conflict that such rivalry engenders? There is no easy answer to this question. What one can say is that in the conflicts between Mazu temples—or Pao-sheng Ta-ti temples or Kuan-kung temples—one does not attempt to physically destroy the enemy. Rather, one tries to win over members of other temples in the same or different cults.

Generally speaking, the rivalry that one witnesses is between the greatest of the island's temples. Such T'ien-hou Kung have the longest and most distinguished history. They also possess direct links to the *tsu-miao* in Fukien or, if not those connections, then links with the ancient Mazu temple on Ma-kong in Peng-hu chain, a temple that claims direct links to the *Meichou tsu-miao*. We have focused on three such temples, the Ch'ao-t'ien Kung in Peikong, the Fon-t'ien Kung in Hsinkong, and the Chen-lan Kung in Tachia.

We have shown that these major temples have created their own networks in Taiwan and have striven to expand these networks.

They have also begun to redevelop ties with the Fukien temples and are attempting to expand these relationships as well as their influence in these ancient *tsu-miao*.

There is one basic rationale for the inter-temple rivalry: it is the need to gain regional and national prestige. These temples are involved in battles for attention and in attempts to make themselves known in the eyes of the Taiwanese public. The expansion of networks the increase in number and size of the almost weekly small-scale pilgrimages, and the steady growth in numbers and of spiritual power of the large-scale festival-related pilgrimages are all facets of this struggle for prestige and for *ling*-spiritual power. The more popular a temple becomes the better it demonstrates the way it adores—it loves and is devoted to the great goddess. This greater degree of adoration—this greater strength of worship, in turn, provides a form of "religious capital" that manifests itself in contributions, in further expansion of temple space and temple influence, and ever more opportunity to praise the goddess in very public ways. This is the kind of religious "capital reinvestment" that is at the very core of Taiwan's—and China's—popular religiosity.

One further point must be made here. These struggles for regional and micro-regional prestige are part of a larger struggle for ethnic power and these temples play a role in the defining of a Taiwanese identity. What I would suggest, following, to an extent, Lin Mei-rung's argument is that to believe in Mazu is to show that one is Taiwanese. As the cult and the various major cult centers have grown, the linkages between Taiwanese ethnic identity and belief in the island's patron goddess have solidified. When one visits a major temple or when one participates in a local or regional pilgrimage—or in a pilgrimage to the *tsu-miao* in Fukien—one feels a part of a body of believers. One also feels that one is participating in a centuries old ritual that reinforces both spiritual sensibility and one's feeling of belonging to the temple, native place, province, and religiocultural region. The older, more traditional Taiwan becomes a part of one's very being in such minutes, hours, and days. It is parallel in certain ways to the feeling a Jew gets while praying in Jerusalem—in *tallis* and *tifillin* at the Wailing Wall. One's sense of religious and ethno-cultural identity is strengthened by such acts of pilgrimage and worship.[39]

These three temples are the same and yet very different. Each temple is dedicated to the service of Mazu. Each of them has felt the blessings of the goddess over the centuries and each has been able to develop and grow and become a major power in the competitive universe of the Taiwanese Mazu temples. By using the medium/message dyads, they have developed means of presenting themselves—their own image of their temple's past, its actual holy space, and its *ling* (efficacy or charisma)—to the larger communities they exist in: local, regional, provincial/national, and trans-strait. By making astute use of the "medium as message" in its three different forms, these temples have developed the means of competing for that most elusive of goals—greater influence in the highly competitive Taiwanese/greater Minnan religious arena.

Notes

1. Hsun Chang, *Incense Offering and Obtaining the Magical Power of Chi: The Mazu (Heavenly Mother) Pilgrimage in Taiwan* (Dissertation for Ph.D. in Anthropology, University of California/Berkeley, 1993).
2. The most useful overviews we have of the Mazu temples in Taiwan is found in the six volumes devoted to the Mazu cult in the *Ch'ien-kuo Fo-cha Tao-kuan T'sung-Lan* (Taipei, 1987). These volumes are part of a larger series prepared and published by a committee made up of representatives of the major temples—text-centered and popular—found in the island. The work is a comprehensive study of the popular tradition as it now exists in Taiwan. It is my opinion that an article or even a short book on the background of this project and on the volumes themselves would be a useful contribution to the literature on religion in Taiwan. The editors take a temple-by-temple approach and have produced profiles of most of the temples devoted to the major gods in the Taiwanese pantheon. However, the editors are also careful to provide the broader context within which one can understand the place of the cult and its temples. Thus, in the first of the volumes on Mazu, there is a basic introduction to the Chinese religious tradition as it now exists on Taiwan. This is followed by a biography of the god or goddess and this in turn is followed by a history of the cult as it evolved in China and in Taiwan. Only then do the editors provide the portraits of the individual temples. These are arranged by county. Each temple is then presented by title in alphabetical (*chu-yin fu-hau*) order.
3. P. Steven Sangren, "Anthropology and Identity in Taiwan: The Relevance of Local Religion" (Conference on Economy and Society in Southeastern China, East Asia Program, Cornell University, 2–3 October 1993. Revised March 1995/September 1995).
4. James L. Watson, "Standardizing the Gods: The Promotion of T'ien-hou (Empress of Heaven)," in David Johnson, Andrew J. Nathan, and Evelyn S. Rawski, eds., *Popular Culture in Late Imperial China*, 292–324. P. Steven Sangren, "History and the Rhetoric of Legitimacy: The Ma Tsu Cult of Taiwan," in *Comparative Studies in Society and History*, vol. 30, no. 4 (October, 1988), 674–697. P. Steven Sangren, "Power and Transcendence in the Mazu Pilgrimages of Taiwan," in *American Anthropologist*, vol. 20, 264–282. Huang Mei-yin, *Taiwan Wen-hua Chi-hwei* (Taipei, 1990).
5. One useful starting point for the study of Mazu is Watson's seminal essay, "Standardizing the Gods," 292–324.

6. The biographies of the Mazu can be found in the pamphlets or large-format works that temples in Fukien or in Taiwan publish. The brief account I have given is based on a fairly elaborate version of what I would term a "mythical biography" that is contained in the T'ainan T'ai-hou Kung's (the T'ainan Great Mother Temple's) booklet. Similar accounts can be found in the booklets (or temple gazeteers) published by the Ch'uanchou T'ien-hou Kung, the Peikong Ch'au-t'ien Kung, and the An-ping T'ien-hou Kung.
7. On popular religion during this period of the Sung dynasty see Valerie Hansen, *Changing Gods in Medieval China 1127–1276* (Princeton: Princeton University Press, 1991). On the history of the goddess and the evolution of her cult see Tsai Hsiang-Chun, *Tai-wan-te Wang-yeh Yu Mazu* (Taipei, 1985).
8. See Watson, "Standardizing the Gods," in Johnson, Nathan, and Rawski, 292–324.
9. On the history of the goddess and the evolution of her cult see Tsai Hsiang-Chun, *Tai-wan-te Wang-yeh Yu Mazu* (Taipei, 1985).
10. On Pao-sheng Ta-ti see Kristopher Schipper, "The Cult of Pao-sheng ta-ti and its spreading to Taiwan: a Case Study of fen-hsiang "in E. B. Vermeer, ed., *Development and Decline of Fukien Province in the 16th and 17th Centuries* (Leiden: E. J. Brill, 1990). See also Kenneth Dean, *Taoist Ritual and Popular Cults of South-east China* (Princeton: Princeton University Press, 1993), 61–97.
11. On this relationship see Chang, "Incense Offering and Obtaining the Magical Power of Chi: The Mazu (Heavenly Mother) Pilgrimage in Taiwan," 66–67.
12. Hsun Chang, "Mazu Hsin-hsiang tzai Liang-an Tsung-chiao Chaio-liou Chung Piao-shih Te T'e-se," *in Liang-an Tsung-chiao Hsian-kuang Yu Chan-wang* (Taipei, 1992), 263–295.
13. This account of the temple's development is based upon material found in the temple's pamphlet as well as material contained in the profile of the temple found in *The Sinorama Guide to Taiwan's Temple/In the Steps of the Pilgrim*, vol. II (Taipei, 1987), 263–267. It is also based upon the detailed account found in *Ch'ien-kuo Fo-cha Tao-kuan T'sung-Lan* (Taipei, 1987), 568.
14. Li Chien-lang, *The Chau-t'ien Temple in Peikang* (Peikong, 1989), 12. *In the Steps of the Pilgrim*, vol. II, 215.
15. Li Chien-lang, *The Chau-t'ien Temple in Peikong*, 19.
16. The origins of the Tachia temple are obscured by the existence of contradictory accounts of the founding of the temple that are found in local and regional gazetteers from such locales as Tamsui and Hsinchu. One such account states that the temple was founded during the final years of the Yung-cheng Era. Another states that the temple was founded in the thirty-fifth year of the Ch'ien-lung Era. A third suggests it was begun in the fiftieth year of the Ch'ien-lung period. There is also a dispute about what the temple's actual name was—whether it was the T'ien-hou Kung or yet another questions concerns whether there were one or two Mazu temples in the town that merged and became the present temple. Questions also exist

concerning which temple the Tachia temple was related to. Some suggest direct links to Meichou while others suggest it was connected to a Mazu temple in nearby Hsinchu. In the 1970s and 1980s scholars writing in major scholarly journals that deal with Taiwan, Taiwan Feng-su and Taiwan Wen-hsuan, examined the local gazetteers and presented their views on the problem of origins. A more recent account can be found in Chang Hsun's dissertation on the Tachia pilgrimage. Dr. Chang carefully avoids the problem of the origins debate and gives a clear picture of temple development in the context of the town's development over time. This allows to show where this temple—one not in a seaport, fits into the larger network of temples to the goddess. See Hsun Chang, "Incense Offering and Obtaining the Magical Power of Chi."

17. Chang Hsun, "Incense Offering and Obtaining the Magical Power of Chi."
18. I have made use of this account because it presents a straightforward narrative of these events that makes comprehensible the various elements discussed in the previous note. It is also a shortened version of the account in the first of the three northern Taiwan volumes of the great encyclopedia, the *Ch'ien-kuo Fo-cha Tao-kuan T'sung-Lan*, an account that captures the flavor of a typical temple gazetteer and takes seriously the supernatural elements that are so much a part of each temple's own self-written history.
19. *Steps of the Pilgrims*, 11.
20 Chang Hsun, "Incense Offering and Obtaining the Magical Power of Chi," 88–89. This account is useful for Ms. Chang shows how this temple developed as the area developed, linking improvements in transportation—meaning the development of rail systems and better road to the process of temple growth.
21. Chang Hsun, "Incense Offering and Obtaining the Magical Power of Chi," 89–91.
22. *Steps of the Pilgrims*, 11.
23. I realize fully well that as I describe these temples and their constituent parts I can introduce, by accident, a subjective impression of what I am seeing. As critics of my earlier paper on temple rivalry have told me, such an impression is not that of those who constructed the temple or who now maintain it. Thus, it represents a distortion of the "space" medium and, in turn, the messages the medium conveys. I have considered this criticism and have tried to rectify the problem in this chapter. Here, my analysis of what I have personally seen and examined is informed by the materials that scholars have written and also materials that the temple committees have written about the temples. My comments are also informed by statements to me by those temple administrators who gave me tours of their temple.
24. In the splendidly produced, well-written, informative and all-but exhaustive volumes in the comprehensive series on Taiwan's religious tradition, *Mazu Temples, Southern Region*, vol. 1, 23–70, there is an introduction on the Chinese religious tradition that includes descriptions of the text-centered—or canonical—traditions and the popular tradition. Included in this systematic article is a description of a typical popular temple. I take this as my model in the description I introduce, and use that model when I examine each of the three temples.

25. *Mazu Temples, Northern Region, vol. 1*, 58.
26. I have based this description on my own visit to and photographs of this temple. The word portraits presented in the *Mazu, Southern Temples, vol. 1* and *Steps of the Pilgrims*, provide me with the detailed vocabulary I lacked and with authoritative details of what I had observed and taken pictures of. The Taiwanese temple is its own universe and it takes time for an outsider to learn what that universe is like. I ask forgiveness for my errors and express my gratitude to the editors of the linked encyclopedias for the mass of detailed data they have presented and for the commentary they have provided the reader.
27. "Fon-t'ien Kung," *Steps of the Pilgrims*, 269. "Fon-t'ien Kung," in *Mazu Temples, Southern Region, vol. I*, 308–321.
28. "Fon-t'ien Kung," *Mazu Temples, Southern Region, vol. I*, 308–321.
29. I have visited the temple thrice. In the spring of 1980, I attended the birthday celebration with hundreds of thousands of pilgrims and visitors and was able to record what I saw on 35 mm slides. I visited the temple again, this time for two leisurely summer days in the summer of 1992. On that second visit, I took more photographs and interviewed one of the temple administrators who gave me a walking tour of the temple. I also obtained the temple publications that I have used in writing my accounts of the temple's history and my description of its appearance. The third time was in May of 2001, when I attended and participated in a conference on the Mazu cult that was sponsored by the Institute of Ethnography, Academia Sinica and the Peikong Fon-t'en Kung Adminsitrative Committee.
30. The three dimensional drawing and the temple's floor plan can be found in Cha Hsiang-chun, editor, *Peikong Ch'au-t'ien Kung Chih* (Peikong: The Peikong Administrative Committee, 1989), 100. See also, "*Mazu Temples*," *Southern Taiwan Region*, vol. 2, 194–213.
31. "Ch'au-t'ien Kung," *Steps of the Pilgrims*, 219.
32. "Ch'au-t'ien Kung," *Steps of the Pilgrims*, 221.
33. "Chen-lang Kung," *Steps of the Pilgrim*, 13. *Northern Taiwan Region, vol. 2*, 330–341. On pilgrimage patterns see *Mazu Temples, Northern Taiwan Region, vol. 3*, 308–327.
34. "Chen-lang Kung," *Steps of the Pilgrims*, 15.
35. I obtained texts from the Taiwanese and mainland Chinese Mazu temples I visited in 1992 and 1993. Among the sites visited on Taiwan were Anp'ing, Annan, Tainan city, Peigong, Hsingong, Changhua, Lukong, and Tachia in Taiwan. In the PRC I visited two major sites in Fukien, Meichou, and Ch'uanchou. Over the years I have also visited the famous Mazu temple in Kowloon. Many types of pamphlets and books were available at the major Taiwanese temples and in the PRC, the Ch'uanchou temple published the most elaborate pamphlet, similar in style to those published in Taiwan. To be sure temples that are devoted to other major Minnan and China-wide gods published their own material, but most are not as elaborate or as extensive as those published by the major Mazu temples. Perhaps the only important exception to this is the Pao-sheng Ta-ti temple in Paichiao. This temple is one of the two rival *tzu-mia* of the Pa-sheng tai-ti cult that are located in Tongan County, near Amoy. This temple, the

Western Temple, has published an elaborate pamphlet, similar in style to those published on Taiwan.

36. Lin Pao-ting, *Anp'ing T'ien-hou Kung Chih* (Anp'ing: T'ien-hou Temple, 1975).
37. On this topic see Murray A. Rubinstein, "Cross the Strait Pilgrimage/Tourism and the Reinvention of the Taiwan/Fujian Popular Religious Matrix." This paper was presented at a conference at Cornell University in 1994 and is available to all interested readers from the author.
38. See the detailed description of the Tachia temple committee in Chang, "Incense Offering and Obtaining the Magical Power of Chi," 93–119.
39. What I have conveyed here is a decidedly personal observation, but one I feel appropriate. The wall of the temple in Jerusalem combines history and almost palpable spirituality that I also feel—even as an outsider in the Meichou *tsu-miao* and even more at the less pretentious *tsu-miao* in Kangli.

Chapter 8

Righteous Brothers and Demon Slayers: Subjectivities and Collective Identities in Taiwanese Temple Processions[1]

Avron A. Boretz

Perhaps no aspect of Taiwanese folk religion is as generally popular, or as controversial as the processions of temple devotees, costumed performance troupes, and entranced spirit mediums that mark the climax of periodic competitive community festivals (*shenming saihui*). Long denigrated as "wasteful" (*langfei*) and "superstitious" (*mixin*) by government officials and government-supported scholars who criticized large-scale community worship and processions as an impediment to economic and political modernity, temple processions have more recently been identified by these same parties (and the mainstream media) as important elements of Taiwanese cultural heritage.[2] With expanding support from private and government sources, "wasteful superstitions" have now become "folk traditions" (*minjian chuantong*) to be documented and preserved for posterity.[3]

Coinciding with the dramatic political transformation of the 1980s, the mainstream print and broadcast media gradually abandoned much of their earlier anti-superstition rhetoric and adopted a more sympathetic tone. In concert with the "Taiwanization" of the KMT and widespread support for "nativization" (*bentuhua*) in all areas of political and cultural life, government rhetoric and policy began to acknowledge the growing popular interest in Taiwanese "folk customs" (*minsu*) in general—and, following Chiang Ching-kuo's visit to Peikang's Chaotian Gong temple in October of 1979, popular religious practice in particular. By the late 1980s, prime-time news programs began to include documentary features on local religious festivals. Significantly, despite a perhaps unavoidable elite bias, most of these reports chose to portray such practices not as the wasteful excesses of uneducated, gullible peasants but as hallmarks of a proud, local cultural tradition.[4] The opening up of travel to the mainland, followed by a number of well-publicized (and occasionally controversial) pilgrimages by Taiwanese temple groups to their ancestral "mother" temples in the mainland, further amplified this enthusiasm.

Astute government and business interests quickly recognized the potential political and economic value inherent in this trend and began to actively promote local culture, including temple processions. Popular religion quickly joined the list of commodifiable cultural properties as official and commercial interests converged. While previously local governments merely tolerated local temple festivals, they now began to act as sponsors, sanctioning them with the telltale title of "Folk Tourism Festival" (*minsu guanguang jie*).

Among the businesses that took early (though not necessarily lucrative) advantage of this new situation were a handful of small, private media companies. Taking their cue from television productions, these small outfits began to make and market their own "docutainment" videos of temple processions and rituals. Many of these small production companies were family-owned, neighborhood-based operations that had previously made nothing more ambitious than wedding and birthday videos. Nevertheless, many of the videos produced by these small outfits were marketed island-wide through department stores and other legitimate distribution networks. Both the performance troupes featured in the videos and also the production companies that made them were thus projected well beyond the confines of their local base of operations. Aside from the potential financial rewards, exposure to a mass audience had important implications both for the practice of ritual performance and also for the identity of the performers. While these new audiences included many who were familiar with the rituals (and more than a few enthusiasts), quite a few videos ended up in the living rooms of young urban families longing to "discover," through such cultural consumption, a heritage and assert an identity lost in the transition to urban modernity. The particular (even idiosyncratic) ritual performances featured in these videos came to represent the entire genre for this particular audience. The troupes whose tapes sold well were now in greater demand; but their styles soon became performance paradigms for troupes throughout the island, even spawning a brief surge of new troupe formation in the early 1990s.

The most obvious and far-reaching effect of this new link with the mass culture market was the exaggerated emphasis on spectacle. These videos emphasized the more theatrical and spectacular aspects of processions, some featuring repeated close-ups of spirit mediums cutting their tongues with saws and piercing their cheeks with steel needles. While self-mutilation has historically been an intrinsic characteristic of Fujianese and Taiwanese spirit medium practice, troupes were now competing to outdo each other in the quantity of blood shed. A reputation for spectacle now meant not only testifying to the power of their patron deity, but commissions and sponsorships as well. Audience expectations changed, and basic modes of representation and participation were, inevitably, adapted. The aesthetics of performance and the training of performers have since undergone an irreversible transformation. As the cases that follow illustrate, however, the most enthusiastic participants tend to downplay the tangible rewards—the expanded financial incentives and promise of notoriety—and emphasize instead their personal obligation as devotees of a deity and responsible members of a collectivity.

The question of the origins of Taiwanese temple processions is beyond the scope of the present chapter, but there is little doubt that Taiwan's temple processions derived from mainland antecedents (de Groot 1982; Katz 1995: 164–165; Dean and Zheng 1993: 152–154). There is, furthermore, ample evidence to show that key characteristics of Taiwan's temple festivals and processions—including the timing of festivals, organization of labor, ritual performance styles (including costuming, postures, and gestures), and the close link between temple festivals/processions and local identity have remained remarkably stable over the last hundred years or so (Lin 1997). Descriptions by Japanese observers documenting ritual practices in the 1930s could easily be applied to today's processions with only the slightest modifications.[5]

Accounting for the persistence and/or the transformation of particular practices, entails a close examination of the links between local, regional, and global events and conditions; and also between collective processes (again, at multiple levels) and what is happening to and for individual protagonists. Here, I intend to make a modest contribution to that project, focusing on microlevel observations and descriptions of a few individual temple procession participants with whom I have worked, extensively but intermittently, since 1989.

A study of identity and subjectivity must begin with the actions and experiences of living persons, and must grapple with the dynamic complexity of psychological and emotional processes and their dialectical relationship to culture and social relations. I have known some of my informants for more than a dozen years, and when feasible, I have tested my observations and reports through discussions and other feedback with these and other informants. I, therefore, submit these records of conversations and descriptions confident that they are sufficiently accurate and precise accounts of actual actions and events. At the same time, they are not meant to be read as character profiles or comprehensive biographies.

As private testimony and public spectacle, processions are simultaneously fields of collective production—of community boundaries, social relations, local identities—and moments of self-creation and self-presentation for individual participants. The success of the spectacle depends both on the actors' individual attributes—talent, experience, and inspiration—and the troupe's level of group training.

For many actors, however, the meaning of the act is not measured by public acclaim or the aesthetics of the performance. Rather, particularly for individuals for whom participation in a temple procession troupe is a central element in their lives, the actions and incidents that take place during a procession—including their own gestures and utterances—are the visible, manifest part of an otherwise-invisible battle between disembodied forces of disorder and decay (ghosts, demons) and the miraculous efficacy of protective deities. In other words, agency is attributed to a spiritual power (a deity) acting through the medium of the actor. In the idiom of Taiwanese popular religion, the actor is chosen by the deity. By this logic, his participation is not a voluntary act. Rather, he is recruited to play an important part in a grand drama. Misrecognizing the source of agency is essential, for otherwise the ritual would be nothing more than amateur theater.

The efficacy of the ritual depends on attributing the action and its progress to unseen but recognizable, powerful, and coherent agents (deities). This efficacy is the work of the ritual: at the level of discourse it is the expected outcome (successful or not) of the action. Again, anticipating and recognizing the outcome presupposes acceptance of at least certain aspects of a particular, shared worldview, in this case a worldview broadly defined and qualified by Taoist cosmology. At the same time, the ritual works to produce the social collectivity and its participants. In fact, the source of the ritual's productive efficacy, in the first place, is their labor in carrying it out. In this sense, through the medium of the ritual, the participants produce themselves, selves that they create *as* ritual producers endowed, paradoxically, with the power of ritual self-creation.

Attributing and projecting agency also authorizes behavior that would otherwise be completely unacceptable, including acts of limited violence and the inversion of hierarchical relationships. As we shall see, this permission greatly expands the range of creative possibilities open to temple procession troupe actors. In the case of military procession troupes, both the deities and the actors who represent them are seen to be potentially violent, socially liminal, and morally ambiguous.

Throughout the mainland as well as in Taiwan, in both Taoist liturgy and popular practice, deities and powers invoked in rituals of exorcism are nearly always martial in character. From the invocation of spirit armies and demon generals in written talismans, to the martial arts poses and weapons of carved deity figurines, to the costumes, gestures, and characterizations of temple procession troupe performers, martial symbols and metaphors are unambiguously associated with exorcistic power.

The particular forms taken by Chinese representations of martial power can be understood as an ideological projection of the violence entailed by the collectivity's production and reproduction of itself in persistently hostile circumstances. The role of stern-countenanced martial deities commanding hordes of spirit soldiers—ritual specialists with their cobra-headed whips, sword-wielding Taoist priests, and spirit mediums piercing their bodies with steel skewers—is to protect a collectivity from the threat of disruption and destruction from forces of chaos and decay: be they in the form of marauding bandits, rival villages, foreign invaders, or malevolent ghosts. Relying on their innate potency and on their learned efficacious techniques (*fa*) using weapons, hand and foot movements, and efficacious words—these men of prowess manifest and embody the necessary powers of material transformation (*ling*).

In southern Taiwan, minor rituals of protection and exorcism, as well as community-wide temple festivals often feature one or another costumed, face-painted, weapon-wielding troupes of young men impersonating martial deities. These fierce, demonic military figures are called *jiang* (H. *chiong*), often translated as "general" but here denoting all field officers and warriors as distinct from the nameless mass of infantry. In the idiom of celestial bureaucracy, these commanders are mostly low-ranking subordinates loyal to one or another of the chief deities. In hagiographies and popular exegesis, the relationship between deities and their *jiang* nearly always follows the Confucian model of complementary reciprocity between lord and minister. As agents responsible for enforcing heaven's will, however, these demigods embody and give a distinct martial character to the efficacious power (*ling*) of the deities.

Responsible for keeping (and occasionally disrupting) order, the *jiang* are the first line of defense against wandering ghosts, malignant demons, and other threats that lurk in the unseen (*wuxing*) layer of everyday reality causing illness, misfortune, and other calamities of personal, family, and community disorder. They are critical to the efficacy of all rituals of purification and exorcism in Taiwanese popular practice. They may be present only in the form of a ritual master's or Taoist priest's invocation and talisman; or they may be embodied, quite literally, by costumed actors who manifest the character, purpose, and power of martial deities in ritual processions.

As the mythobiographies of *jiang* attest, the desired talents of a *jiang* are indeed those of a warrior. Martial skills, fierce aggression, and arrogance are all typical characteristics of the demons—the wandering ghosts of righteous men cut down violently in the prime of life (often a group of sworn brothers who share the same fate)—who are eventually transformed into *jiang*. This transformation is enacted through encounters between these wandering spirits and orthodox deities. The magic battles that follow always end with the now-subdued and awed demons pledging eternal loyalty to the deity.

There is a revealing paradox implicit in these myths: as loyal subordinates, the *jiang* conform perfectly to the Confucian ideal. But although the *jiang* are said to be transformed through their adoption, in fact they are merely domesticated demons. It is their origins as violent, capricious, and potentially destructive demons that makes them useful as enforcers and bodyguards. When the situation requires it, the master deity can unleash this demonic, destructive power. Yet, the *jiang*'s masters fear them even as they control them and exploit their power. They must be watched constantly and carefully controlled, for their *yin* demonic power can overwhelm even their superior *yang* masters.[6] I would argue that the attraction of the *jiang* for the young men who impersonate them, in fact, derives from this very contradiction. Like these young men, who are sons and apprentices in a society dominated by older men, the *jiang* are subordinates. But sons and apprentices are patriarchs in waiting, as it were. The desire for autonomy, even domination, is vital to the construction of masculine identity in Taiwan, and it is a commonplace in Chinese and Taiwanese society that impatience and frustration fueled by their desire to reach the top may drive subordinates to resist and even attempt to invert the hierarchy (Sangren 1984). This tension is, thus, clearly a projection of the vexed relationship between older and younger men (including, but not limited to the father–son relationship) in Taiwanese society.

In Taiwan, local identity and solidarity are still often expressed in terms of the exclusion of outsiders enforced by threats of violence and retribution. One informant from the village of Beinan once told me:

> The kids in Beinan stick together. No one dares to come into the village to make trouble. If someone has a vendetta, or whatever, and comes into the village to look for someone, the guy being threatened just gives a shout and everyone comes running to back him up. One time, some gangsters from Fengtian beat up a Beinan kid and smashed his motorcycle. It turned out to be a case of mistaken identity. The head of the Fengtian village council led a whole delegation to Beinan to personally apologize for the incident, a real loss of face. But they knew that if they didn't do this, there would be a bloody vendetta between the two villages and they knew they'd lose and people would get hurt, maybe killed. People in Beinan have a saying that goes, "If you come into Beinan and spit one gob of saliva, we will drown you." (*Li na jip Pi-lam chng, chit-e lang phui chit-e chhui-noa, to e ka li im-si.*) [That is, any outside threat entails an overwhelming attack in kind on the intruder.]

This ethos may very well have grown out of the violent uncertainties of premodern Taiwanese frontier life; in any event, the very same uncertainties and undertones of violence still pervade the lives of those on the economic and social margins. In Beinan and other towns and villages throughout Taiwan, the young "braves" (*yong, iong*) who band together in defense of the village when territory or honor are at stake are often the same "gangsters" who extort protection money from local businesses, run gambling, prostitution, and smuggling rackets, and control local politics, enforcing control through intimidation and violence. The ritual importance of Taiwan's military temple procession troupes derives, in part, from their identification with local self-protection from at least the nineteenth century through the early 1960s (Sutton 1996, Jordan 1972, and the like). Indeed, throughout their history in Taiwan, the troupes have not only represented local solidarity and a collective response to the threat of intrusion; they have been part and parcel of the organization of local social and political power. Documentary and anecdotal evidence suggests that at least two of the most popular types of Taiwanese military temple troupe—the Military Retainers and the Song Jiang Battalions—can be traced to Qing banner troupes from Fuzhou and local militia organizations, respectively (Sutton 1996: 216; Jordan 1972: 48–52). More importantly, however, is the association of supernatural power with the practical power of an organized, trained, collective armed force. Moreover, whatever their (probably very limited) practical value as militias, the troupes manifest (indeed, are the means and medium of) the organizational abilities and military prowess of a locality.

As a local event in Taiwan's least accessible and least developed area, the Taitung Lantern Festival attracted little outside attention until the late 1980s. A rail link to Kaohsiung (opened in 1992), airport expansion, highway improvement, creation of the Eastern Taiwan National Seashore, and considerable outside investment in hotel, restaurant, and entertainment ventures helped realize the local government's vision of tourism-based economic expansion during the 1980s and 1990s. Indeed, tourists from Taipei, Kaohsiung, and other west-coast cities have been flocking to Taitung in increasing numbers since the late 1980s to view the spectacle of the Lantern Festival and enjoy Taitung's pristine mountains and seacoast. As one would expect, this new popularity has placed unprecedented pressure on the natural and social environment.[7]

For two days every year, starting on the day before the first full moon of the lunar calendar, the streets of Taitung are filled with procession troupes, entranced spirit mediums, and throngs of onlookers. By the end of the second day, the streets of the downtown area are ankle-deep in firecracker paper, and the air is thick with smoke and the smell of gunpowder. All along Taitung's streets and alleys, businesses and households put out small tables laden with sacrificial offerings to the deities and their celestial armies.

At the height of the procession, at many of these altars and at the gates of the scores of temples that dot the town, every few minutes a temple group passes and enacts a short ritual. At the vanguard of these temple procession groups is a small wooden or rattan sedan chair (*jiaozi, kio-a*) mounted on two horizontal bamboo or wood poles and handled by a team of two or four young men. As they approach, they break into a run, and as the residents or proprietors or temple devotees throw packets of exploding firecrackers at their feet, they stop and dip the chair either one or three times[8] before rushing off to rejoin the parade. This practice is an essential part of any temple procession or pilgrimage everywhere in Taiwan throughout the ritual year, but it takes on a particular intensity in this Lantern Festival procession.[9]

Following the *kio-a* team and preceding the main deity's palanquin, at least for those temples with the means and manpower to maintain them, are the performance troupes, or *zhentou* (H. *tin-thau*).[10] Temple procession performance troupes have been an integral part of popular religious ritual for centuries, if not longer. Costume styles, choreography, deity characterizations, and the blurring of the line between theatrical and shamanic performances link Taiwan's contemporary temple festival processions with similar practices described much earlier in southeastern China. At the same time, local innovation and adaptation, particularly since communication between Taiwan and the mainland were stifled after 1895, have had a profound influence on the form and practice of ritual performances. During the late nineteenth and through the twentieth century, local creativity and competition produced a confusing profusion of new troupe types and variations and more flamboyant performance styles (Sutton 1996: 218–219). New and even more extravagant variations continue to proliferate under the pressure of mass exposure through television and other media. While traditionally performance troupes were local, ad hoc, and often transitory groups of volunteers identified closely with a particular community or territorial cult (Lin 1997), the higher stakes entailed by access to mass audiences and commercial opportunities (including government-sponsored "folk festival" performances at the Chiang Kaishek Memorial in Taipei) have stimulated a trend toward professionalization. Broadly speaking, *zhentou* are generally divided into two main categories: civil troupes (*wenzhen*; H. *bun-tin*) and martial troupes (*wuzhen*; H. *bu-tin*).[11]

With the exception of the large "frame puppets" (H. *tai-sin ang-a*) so visible in the midst of processions, the martial troupes are made up of actors costumed and made up in the style of Chinese stage opera. As befits warriors and military officials, they carry a variety of weapons and implements used in the capture, intimidation, torture, punishment, and dispatching of wandering ghosts. Some are accurate replicas of swords, spears, and halberds or traditional tools of capture, restraint, and torture, while others are fanciful spirit weapons or symbolic items like rubber snakes and flower baskets. They march along the procession route using the Taoist "dipper step" (*qi xing bu*), occasionally forming battle arrays through a sequence of stances and gestures derived from traditional martial stage techniques. A few troupes, notably the lion dancers and Song Jiang Battallions are often offshoots of martial arts schools and include demonstrations of more practical martial skills in their performance repertoires (Lin 1997; Boretz 2001).

In addition to the *kio-a* teams, the procession troupes, and the main palanquin, nearly every temple group will have among its numbers one or more entranced shamans (*jitong, tangki*). Stripped to the waist, wearing an embroidered vest and/or a red sash tied at the waist, these *tangki* also wield "precious magic tools," but in this case, they are real weapons used for self-mortification (to verify the presence of the deity, as well as to demonstrate the deity's awesome exorcistic power), and to capture and slay demons directly. Many protagonists move easily between these two kinds of roles throughout a single procession, if it lasts long enough, or at one time or another during the course of their active participation in procession rituals.

One of the most popular temple procession troupes is the formation known as the Eight (sometimes Ten) Military Retainers (*ba/shi jia jiang*; H. *pat/sip ka chiong*).[12] For most of the duration of most rituals in which they take part, the Retainers simply perform steps and other prescribed ritual acts of exorcism and purification, carrying out the directives of their commander (usually identified as the martial adept Liu Wenda, third of the Emperors of the Five Blessings).[13] Occasionally, however, Military Retainers may be directly possessed by their patron deities, going into a sometimes-violent trance similar to those of Taiwanese spirit mediums (H. *tangki*).[14] Like *tangki*, Military Retainers troupe members must abstain from sex and avoid certain foods (especially beef) for at least three days prior to the ritual. But unlike tangki, who are understood to embody divinity only when in the throes of possession trance, the Retainers become fused with their patron spirits through a ritual of separation, remaining sacralized (i.e., ritually off-limits to all but the temple's own ritual assistants and senior members) throughout the procession.

Divination establishes a karmic bond between the deity and the Military Retainer actor.[15] Just by having once been encostumed as a Retainer, one gains the protection of a supernatural patron. The mask of face-paint of that particular deity, for instance, is said to inhere on the inner surface of the face of the actor for the rest of his life.[16] The actors, then, are "possessed" by the deities whom they represent as soon as they are costumed and ritually separated. While Retainers may be possessed during processions, trance is, therefore, not a necessary mark of sacred presence. The identification of the actor and the deity is enhanced, of course, if the actor shows proficiency in the ritual moves and becomes possessed at critical moments.

I present three case studies documenting the procession troupe performances of individuals whose identities are closely bound up with temple procession ritual performance. All have played many ritual roles at one time or another, but at the point I joined their activities, each had settled into a preferred mode of action and exegesis. Each represents a different aspect of ritual procession performance, as well as a different aspect of the social and economic production connected to these ritual roles, but all three identify,[17] in different modes, with Wang Ye, the Eight Military Retainers, martial arts, the underground economy, and violent, dramatic spirit possession.

Two of these informants (Pik-khut-a and A-kiat-a) are natives of Taitung who migrated to the urban areas of the western plain several years before I met them (at which time one was in his late 30s, the other in his early 40s, close in age but actually belonging to different cohorts). They come back to Taitung every year at Yüanxiao to participate in Zhonghe Temple's procession activities, but their active involvement in ritual and performance is year-round and more closely tied to temples where they work and reside. The third, A-peh, is a native of Kaohsiung County who moved to the outskirts of Taichung City and joined Pik-khut-a's shrine as a spirit medium and master trainer of a troupe of professional[18] Military Retainers performers.

Pik-khut-a[19] is the "Hall Master" (*tangzhu, thng-chu*) of a small storefront Wang Ye temple in the suburbs of Taichung. Although this temple was "divided off" from a temple near Kaohsiung devoted to Wenfu Qiansui, Pik-khut-a brings his retinue of "military retainers" and "street performers"[20] back to Taitung each year to join Zhonghe Temple's Yüanxiao procession.

Pik-khut-a, like many working-class Taiwanese men, is a member of a sworn brotherhood. Also like many working-class Taiwanese men, at some point in his teens or twenties he was a member of an organized gang, a fact attested to by the dragon tattoos on his arms and back. In Chinese mythohistory, the bond of sworn brotherhood is usually a prerequisite for manly heroism. Modeled after the paradigmatic Peach Garden blood oath described in the classical novel *Sanguo Yanyi*, the oath of brotherhood establishes a hierarchical structure of absolute solidarity. This oath articulates an important axiom of Chinese masculinity, namely, that being a man is first and foremost a matter of identifying with a collectivity, a group of male peers. These informal groups, while lacking the jural legitimacy of patriclan halls, for instance, are nonetheless among the most basic practical units of Chinese (and Taiwanese) society.[21] For sworn brothers this oath entails absolute mutual allegiance that transcends all other social obligations, and generosity without regard to financial capacity. Among the unwritten rules of this cultural mode is that, while "all men are brothers," brotherhood is a hierarchical order based nominally on seniority but in practice depends as much, or more, on competition and negotiation. Thus, written into the compact of friendship among sworn brothers[22] is an imperative to constantly challenge and test each others' generosity, loyalty, and appetites. In modern Taiwan, the most popular venues for such competition are restaurants, KTVs, and festival banquets.

It is not just the ability to hold one's liquor that is at stake in these almost nightly competitions, but the willingness to continue drinking at your hosts' and friends' invitation even after going far beyond your limit. Having a "good time" is not the point. In the fluid dynamics of an evening out with one's "brothers" (*xiongdi, hia^{n}-ti*) and associates, microhierarchies are subtly affirmed or rearranged. Business is often the subtext and sometimes the pretext of these evenings, but is rarely discussed at these times, particularly when it involves activities on or beyond the verge of legality. As with his own sworn siblings, the host seeks to build his relationships with his clients and potential clients by establishing himself as a man of heroic ways, an exemplar of "loyalty and reciprocity." Thus, the heroic, martial archetypes are inscribed within the most basic social interactions and phenomenological processes across a wide range of social groups and activities. The martial ethos pervades and defines the cultural production of male subjects in Taiwanese working-class society.

This ethos, defined and explicated through the narratives and representations of the martial, implicates most social activity in the development of this male identity. One day in May 1991, Pik-khut-a came up to Taipei to help out with the procession activities at Gongrong Temple during the Mazu procession in Hsintien, Taipei County.[23] When he arrived, accompanied by an assistant, it was clear he was still suffering the aftermath of the previous night's carousing. Late in the morning, as we walked together in ninety-five-degree heat behind the god's sedan chair, Pik-khut-a explained "Last night we were drinking XO, that client is really over-zealous. Liu-bu [the patron deity of the Ten Military Retainers] got me, he grabbed me. I really fear getting possessed by Liu-bu, he made me throw it all up. Whoo, I'm exhausted today." "Liu-bu got hold of you?" I asked. "Yup. I threw the whole thing up." "You've still got energy and spirit today!" "Well, I have to, that's the way it has to be . . ."

The procession went on from dawn until late in the evening of a clear, hot, dry early summer day. Pik-khut-a worked especially hard carrying the "tools of interrogation and punishment" (*xingju, heng-ku*) to lead the Ten Military Retainers through their gruelingly repetitive house-exorcising rituals all along the route. The role requires marching in step, carrying a bamboo or wooden yoke laden with miniature replicas of the implements of torture found in the late imperial yamen. At ritually important points along the procession route, and at the houses of supplicants who request the exorcism of their premises, he leads the Military Retainers through a series of dramatic ritual actions. Doing this properly and efficaciously requires a long series of low horse- and bow-stance (*mabu, gongbu*) steps. This entails martial arts expertise, strength, flexibility, and endurance, and Pik-khut-a, with his extensive martial arts background, is one of the best I have seen performing this role. Later that day, as the procession neared the intersection of two main roads, Pik-khut-a, nearly exhausted by his exertions was suddenly and violently possessed by his patron deity, Wenfu Qiansui.[24]

Well after nightfall the group returned to the home temple. After drinking far more than required at the celebratory feast that followed, Pik-khut-a stood up and announced his departure for Taichung. As he prepared to leave, one of the temple committee representatives offered him a "red envelope" filled with NT$10,000. Laughing and appearing surprised, he refused to accept the money, pushing the envelope away repeatedly with a show of distaste.

The dynamics of this exchange are complex. Pik-khut-a's livelihood actually depended on the demonstrated supernatural efficacy of the deity in his storefront shrine, and he maintained a professional Military Retainers troupe that hired out for cash without apology. However, as his frequent testimonies to the miraculous power of the deity attested, he strongly believed that his own fortunes were bound up with Wenfu Qiansui (often referred to with the honorific Wangye), his patron god. According to Pik-khut-a, his success or failure in business, which at the time included both wholesale soft drink distribution and gambling, depended entirely on the graces of the deity. At the same time, he pointed out (presumably without irony) Wangye's power depended in turn on the temple's notoriety and on the number of devotees who came to burn incense, enter into reciprocal obligation with the deity, and join the temple's pilgrimages and processions. Pik-khut-a had been very successful at drawing a strong and consistent following into his storefront cult, and his Military Retainers and spirit mediums, who enjoyed a growing reputation for spectacle, were certainly a point of attraction for many believers. Involved as a teenager in a more insidious form of brotherhood, his upper body bore the trademark dragon tattoos. His business activities and connections extended in many directions and occasionally crossed the lines of legality, although he was decidedly not engaged in any violent or coercive criminality.

All this was well known to the Gongrong Temple committee members, who saw themselves, in contrast to the "professional" religionists like Pik-khut-a, as humble, devout sacrificers. They were mostly small businessmen engaged in leather-goods manufacturing, and, as they liked to emphasize, gained no financial profit from their religious exertions. Quite the contrary, they put up their own funds to sponsor feasts, processions, and other activities, and took no outside contributions, not even for successful cures rendered through shamanic healing sessions.

Pik-khut-a had to work hard to defuse his negative public image. He had come, he wanted the Gongrong Temple group to understand, through an obligation of service to the deity. It was an implicit overture of friendship and cooperation between his temple and theirs. His refusal of cash compensation for his labor obligated the Gongrong Temple group to payment in kind (i.e., lending a hand and giving face) sometime in the future. Their reluctance to enter into such a bargain suggests that they might have been somewhat wary of Pik-khut-a and his gangster reputation. By firmly and absolutely rejecting the offer of compensation, Pik-khut-a demonstrated, to himself and the Gongrong group, that he was, after all, a righteous man and a devoted servant of the deity, not a hoodlum trying to exploit Xingfu Qiansui's good reputation.

Pik-khut-a's cavalier refusal of the money can be seen as an act through which he defined himself as an individual person as well as a social subject. Without guile—for I never detected any cynicism or conscious deceit in his self-presentation—Pik-khut-a strove to be a man of integrity in the tradition of the Chinese knight-errant. His refusal to accept payment defined him as a righteous brother, but it also marked a trajectory of desire that obviously diverged from the Gongrong Temple group. The offer, followed by the refusal affirmed, rather than deflected, the group's view of him as an outsider and member of a the criminal underworld and of themselves as upwardly-mobile but honest businessmen. While each party had its own perspective, this interchange demonstrated a collective subjectivity constituted, in part, by a shared set of mythohistorical narratives.

A-peh was, for a time, the coach and master teacher of the Military Retainers troupe at Pik-khut-a's temple. He also trained the "street troupe" of self-mortifiers, and was the chief *tangki* for Wenfu Qiansui, the temple's patron deity. Young, charismatic, proficient in martial arts, and recently married, A-peh seemed full of self-confidence and social savvy. While Pik-khut-a handled the everyday affairs of the temple and did business with associates of his own generation, A-peh organized his own subgroup within the temple, a group of about twenty sworn brothers in their late teens and early twenties. Most of them were gang runners, small-time toughs known colloquially as "bamboo partridges" (*tek-ke-a*). All of them had obligations to other brotherhoods and gangs that sometimes came into conflict with their responsibilities as temple troupe members, and there were occasions when Pik-khut-a and A-peh found it difficult to assemble the full complement of Military Retainers and self-mortifiers. The first time I met A-peh was just such an occasion.

I had arrived at the temple at dawn, as preparations were already underway for the day's activities. Pik-khut-a's troupe had been "invited" (hired) by the local Mazu temple to join their pilgrimage procession, a major event that was to take us to Chu-nan, Tou-liu, and then back to Taichung all in one day. Arrangements had been made well in advance, and much face (and money) was at stake. At 7:30, as time for departure neared, two of the troupe members had still not shown up. When they finally appeared, exhausted and apparently suffering from amphetamine withdrawal symptoms, A-peh blew up. He punched out a plate glass window, showering one of the troupe members—an innocent party, as it turned out—with glass shards. His face cut and bleeding, this unfortunate individual offered no complaint, but went quietly upstairs to have his cuts treated by Pik-khut-a's wife. Pik-khut-a looked unconcerned by the violent outburst and its unfortunate outcome: "It doesn't matter," he explained matter-of-factly, "they're sworn brothers."

Having finally calmed down, A-peh went upstairs to begin the laborious process of painting the faces of the Military Retainers. From the Civil and Military Messengers at the head of the column to the Four Seasons and the Civil and Military Magistrates who bring up the rear, each role, with the exception of Heng-ku Ia, has its unique facepainting pattern and costume. The styles of adornment and the techniques of painting are derived directly from traditional opera. The intentionality, however, is obviously quite different.

Fully decorated and attired, the Retainers took their place at the long table set up in front of the main altar. Seated on packets of spirit money set on plastic stools, their ritual implements arrayed on the table before them, the troupe sat silently in formation. Pik-khut-a, A-piang, and some others began to chant the mantra of the Ten Great Deities (the Retainers), accompanied by drums and gongs. After about twenty minutes, two of the Retainers became possessed, and had to be exorcised. Finally, the divination blocks were thrown to inquire whether or not the deities had in fact descended. When their presence was confirmed by three consecutive positive throws, the chant came to an end. Pik-khut-a then read the petition, asking for the assistance of Liubu Dadi and his contingent of Retainers for carrying out the tasks that lay before us that day. A talisman was burned and the ashes crumbled into a bowl of water, and each Retainer drank in turn. Each pair, beginning with the Messengers and continuing back to the high-ranking Magistrates, came to the front and worshiped with incense, recreating the Retainers' oath of allegiance to Liubu Dadi. The ritual of obeisance complete, the Retainers were commanded to stand again. At this command, the actors now embody the deities, without the usual progression into possession trance. Passing their implements of exorcism over the main censer as they exited, they filed out of the temple in pairs. Crossing a burning pyre of spirit money, the Retainers fell into formation and began their ritual steps, making a circuit of the alley and returning to make a final bow of obeisance to Wenfu Qiansui before setting out on the day's appointed mission.

After a two-hour ride in the back of a flatbed truck, we reached Chu-nan. Debarking, the initial rituals of setting up formation and presenting arms was repeated, with the image of Liubu set on a cart substituting for the temple altar. Then the long, slow procession to the Chu-nan Mazu Temple began. Since this was a pilgrimage on "foreign" territory, for a temple affiliated with a different lineage and a different deity, there were no house or business exorcisms. A few "spontaneous" possessions took place at crossroads and other places of ritual danger, but not until the troupe—now part of a long, snaking parade—approached within a few blocks of the temple did the number and intensity of possessions start to peak. It was at the point where the CTV cameras first appeared that A-peh became possessed. This was, it turned out, an officially designated "Tourism Folk Festival," and the media were out in force.

The possession started when A-peh started to stare vacantly ahead of him, immobile. Soon he began to shake, and Pik-khut-a and others rushed over to remove his shirt and tie a strip of red cloth around his waist. Suddenly, A-peh broke into a crisp, fast, and powerful martial arts form—Wenfu Qiansui was a military commander and failed military exam candidate (*wujüren*), according to the mythohistory. Since this indicated the presence of the deity, the censer was brought over, and Wenfu Qiansui was offered incense. A-peh/Wenfu grabbed the censer and started to eat the burning incense, smearing his face and torso until he was entirely black with incense ash. He tied another red cloth around his head, tying tubes of paper talismans upright at the corners. After marching a few dipper steps further, A-peh/Wenfu stopped again and called for a long bronze needle. At this point the black flag, the talisman of the Lushan Taoist lineage of black cloth with white characters, was used to conceal A-peh's face. With Pik-khut-a's assistance, the needle was carefully inserted through A-peh's left cheek.

Inserting steel needles through the cheek and tongue are time-honored techniques of self-mortification among Chinese spirit mediums.[25] With the cameras rolling, A-peh, however, did not stop at simply piercing the cheek. He took the needle and bent it into a loop, and stuck the sharp end again through his right cheek. He then called for two spike balls, which he had fastened to his heels. Finally, he called for his Dipper Sword (*qixing baojian*). Thus equipped, he alternately walked in ritual step and rode the god's palanquin, gesturing commands to stop and start the procession as he deemed appropriate.

The Military Retainers practice a variety of efficacious formations and drills, called by such names as "Eight Trigrams Formation" (*bagua zhen*), "Lotus Bloom Formation" (*lianhua zhen*), and "Opening the Four Gates" (*kai simen*). Unlike the exorcism of households (*zhen zhai, chin theh*), these formations are usually only performed at moments of ritual climax, such as when entering the precincts of a major temple. The Mazu Temple courtyard was packed with devotees, tourists, and news cameramen. The Military Retainers, who are first and foremost the private enforcers and bodyguards of Wang Ye, entered the courtyard first and began a long, energetically performed drill formation—"Opening the Four Gates" (*kai si men*). On this day, the performance went on for almost twenty minutes as the troupe elaborated the traditional moves with acrobatic spins, acted out the story of Seventh and Eighth Elders,[26] and fell one by one into trance as the cameras rolled.

Having paid their respects to Mazu at the Jade Emperor's Censer (*tiangong lu*), the Retainers doubled back and set up a gauntlet, lining up at the sides of the courtyard to guard at the entrance of Wenfu Qiansui and Liubu Dadi onto the scene. Preceding the palanquin and A-peh were five or six of the other *tangki*. Stripped to the waist, they began beating themselves on the back and forehead with nail clubs (*ding gun, teng kun*), spike balls, and other traditional and nontraditional implements of self-mortification (*fabao, hoat-po*). A-hiang began to cut his tongue with a saw, and Han-chi-a sliced his tongue with an ax blade. The cameras captured all this from every angle.

Finally A-peh entered, spike balls still firmly bound to his bare feet. As the din of firecrackers began a long crescendo, he strode up (still walking the dipper step) to meet the host temple's entranced *tangki*, who had come forward to greet the arriving deity. The two communed animatedly for a minute or so. A-peh then called for a nail club, and began to spike himself in the back over and over again. Finally, the two entranced shamans turned and marched, side by side past the main censer and into the temple. The crowd—and the cameras—followed. Inside, A-peh went up to the main altar. He started pounding and writing with his fist on a table set up for just this purpose. Pik-khut-a and a representative of the Mazu congregation stood by, interpreting the god's written communication.

A-peh then stood back, and removed the spike balls from his feet, and the metal rod from his mouth. He stood up on the altar table, gave out a shriek, somersaulted backwards and collapsed into the arms of two waiting helpers. This signaled the departure of the deity and the end of the possession. Clips of A-peh in trance and the Military Retainers' performance were included in that evening's local evening television news, which featured an extended report on the day's events at the temple.

I saw A-peh in trance several times over a period of three years. While he occasionally added or deleted one or another detail, the "narrative" and style of his trance performance was remarkably uniform. His transition always began with convulsions, violent face contortions, and foaming at the mouth, followed immediately by a fierce, perfectly coordinated, martial arts set. At this point assistants would rush over to remove his shirt, tie a red bandanna around his head and the *tangki*'s apron around his waist. As he was fed sandalwood incense smoke he would invariably grab handfuls of burning incense and smear his face with soot. He would then call for his trademark copper rod and insert it in his cheeks himself; and have spike balls tied to the soles of his feet. He would almost always alternate walking with riding on the palanquin, and his trances usually ended with a backwards somersault off the altar. Unlike Pik-khut-a and many of the troupe members, A-peh remained aloof and silent whenever I tried to discuss his trances and his career with him, and I observed that he rarely joined in discussions of religious practice or cosmology with his older colleagues. But in his coaching of the Military Retainers, his demand for allegiance from the group, his "righteous" response to drinking challenges and toasts at feasts, and through his spectacular trance performances, he showed himself to be an adept and confident handler of people and a savvy manipulator of public attention. Hardly shy or socially inept, then, A-peh used silence to create an aura of mystery around his entranced self. This mystery enhanced the authenticity of his performance; yet, paradoxically, it also served to fuse A-peh's everyday self and his trance persona into a recognizably consistent character.

Among those who partake in this feast of propitiation is the procession troupe popularly called the Eight Military Retainers. Beginning in 1990, I was honored with the privilege of participating in one of Taitung's best-known Military Retainers troupes, playing the role of the Civil Magistrate. On many occasions, I was "working" opposite A-kiat-a, who soon became one of my most enthusiastic informants. A-kiat-a's identification with one role in the Military Retainers troupe (the Military Magistrate) has defined his public persona for more than twenty-five years. While many of his peers have become embroiled in the wranglings and power intrigues of temple politics, A-kiat-a's quiet, self-effacing manner have enabled him to remain disengaged. For most of the procession, the Military Magistrate, like A-kiat-a himself, stays mostly in the background. But the Military Magistrate is an enforcer—and as such, is expected to act forcefully and demonstratively when the situation calls for it.

During the first night of the 1990 Yüanxiao procession, as the Military Retainers reached a storefront shrine along Zhonghua Road, A-kiat-a seemed agitated. A ritualist carrying a censer came out of the shrine to greet the troupe and offer incense to the Retainers.[27] As the front ranks finished imbibing their share of incense smoke and proceeded on to acknowledge the shrine's deity, A-kiat-a, marching in place, fell into a full-fledged trance. When his turn came, A-kiat-a, his head shaking rhythmically from side to side, his face fixed in the magisterial frown of a stern and angry deity, leapt forward, grabbed the censer, ate several mouthfuls of burning incense, and kicked the censer-bearer to the ground. As A-kiat-a coiled into an attack posture and hovered threateningly over the still-supine ritualist, the possessed spirit-medium of this temple came out and, moving slowly through the routines of a martial arts form, offered half-hearted, symbolic resistance. Within a few seconds, a friend of A-kiat-a's, a ritual specialist from Taichung, jumped in and recited a spell. As he channeled his spell-empowered qi through a two-fingered thrust of his right hand, A-kiat-a fell back into the arms of two waiting assistants, and withdrew from the trance. A hurried explanation was offered the offended temple's representatives, the incident passed, and the procession moved on.

Inscribed within both prescribed ritual and the spontaneous, "in-character" possession modes of Military Retainers performance, we can see a competitive, violent side of Taiwanese social reality. When a Retainers troupe encounters a similar troupe from another temple, for instance, each member must hide his face, for meeting the gaze of his counterpart in the other troupe would compel him to fight. Although real fights rarely occur, one can often see troupe members become suddenly possessed and wildly agitated when such an encounter happens. The Military Retainers, my informants often enjoyed telling me *sotto voce*, are "celestial liumang," enforcers and hit men working for Wangye and the Plague Gods (Wufu Dadi). Stories of fistfights between rival temples in years past are thus part of the lore of every established Military Retainers troupe in Taitung.

Nevertheless, this was the first time I had witnessed an act of real violence in the context of ritual, and naturally I was anxious to discover the cause of the incident. In conversations a few days later with A-gi and other temple regulars, I learned that the storefront temple on Zhonghua Road had been founded about ten years earlier by a former member of the temple who had departed angrily over some unspecified disagreement. A-kiat-a's violent rebuff was clearly intended to reprimand the renegade temple for their transgression and make it clear that once they left the fold they had lost the sanction of the deity and had no further access to his numinous power or to the advantages of belonging to his fold, including the security of alliance with the mother temple group.

This performance was clearly meant for those few temple elders for whom the schism would have significance, most of whom were present that evening. Few of the younger members of the troupe were aware that the two temples had a history of antagonism. Yet, A-kiat-a's outburst was equally meaningful for them, but in a very different way. As soon as he struck the incense bearer, the other members of the troupe responded immediately by halting their forward progress and closing ranks, ready for action if necessary. What that action might be no one would say, but another incident involving A-kiat-a revealed one possibility.

The long-established Heavenly Official Temple is one of the few temples in Taitung that can be described as a "territorial cult." Nearly all the devotees of this temple are residents of the neighborhoods immediately adjacent to the temple, and both the temple leadership and its procession performers are comprised of local young men—among them many "toughs" and gang members—hailing from these neighborhoods. A-kiat-a's temple, although its congregation is more diverse than the Heavenly Official Temple's, was also associated with one group of "sworn brothers" for most of the 1950s and 1960s. Although that group is now largely dispersed, the hostile, competitive relationship engendered by those "turf wars" has become a reified given of both temples' "processional identities."

Found in Chenghuang (City God, or God of Walls and Moats) temples in Taiwan, the generals Xie and Fan—Seventh and Eighth Masters—serve Chenghuang by escorting across the Naihe Bridge the souls of the recently deceased and any wandering ghosts they may capture. The highlight of the Heavenly Official Temple's procession is the performance of the two large wood-frame effigies of these fearsome demigods.

On the morning of the second day of the 1991 Yüanxiao procession, the Military Retainers had been invited to purify a small dry-goods shop just down the street from the Heavenly Official Temple. As it happened, the Heavenly Official Temple's procession troupe was only starting their morning rounds and had just exited the temple gate and turned the corner. At that moment, most of the Military Retainers troupe had already entered the shop and were in the process of purifying the premises. In this ritual, called "securing [lit. subduing] the residence" (*zhen zhai*, H. *chin-theh*) the Civil and Military Magistrates do not enter the premises, but stand guard outside. A-kiat-a was, thus, in position to see the Heavenly Official Temple's Eighth Master as the other group approached. He became possessed suddenly, and striding angrily toward Eighth Master, he confronted the wood-frame demon, shaking his feather fan threateningly in its face. As the rest of the troupe finished their purification and exited the shop, they saw the commotion and headed into the fray. Seeing the situation getting out of hand, procession leaders from both groups ran in to separate the troupes. After applying a talisman to bring A-kiat-a out of his trance, some of our own group guided the now-listless A-kiat-a back toward the shop, while others exchanged greetings and explanations with the Heavenly Official Temple leaders. The procession disappeared down the street, and A-kiat-a was released from his trance.

In both of these incidents, A-kiat-a's violent trance performance was a declaration of his temple's place in the local temple hierarchy, an assertion of group cohesion in the face of social disintegration and intergroup rivalry. Yet, it was also a creative and productive act emanating from and reflecting upon one individual. In the first case, A-kiat-a was performing primarily for the inner circle in a situation whose outcome was hardly in doubt. The storefront temple would have been at a severe disadvantage, both in terms of manpower and the moral assessment of the community. There was little real risk, although had anyone in the other group been moved to rage in response to the humiliation, the situation could have degenerated quickly. In the second incident, however, the other group actually outnumbered our own; the two groups had a history of violent confrontation that included a number of armed brawls; and we were on the other group's turf. A-kiat-a's challenge in this case was truly a risky move, and yet it had perhaps even more value just for that reason. As some temple insiders (including some of the troupe members) later described it, terrified and overwhelmed, the Heavenly Official Temple's Eighth Master acknowledged A-kiat-a's Military Magistrate as his teacher, affirming, at least in this interpretation of events, the superior position of Zhonghe Temple. Were A-kiat-a acting on his own, this would have been a foolhardy show of bravado; but understood as the actions of a powerful and determined deity, temple members saw it instead as a higher-order sanctioning of their own version of the local temple hierarchy. The fact of possession testifies to A-kiat-a's special relationship to the temple's supernatural patrons—to Wang Ye, and to his own character, the Military Magistrate. The contents of his trances, furthermore, allow him to create a special role for himself within the group, that of a qualified arbiter of social and moral issues pertaining to temple mythohistory. In both incidents, A-kiat-a was stepping up to fight for his group, but he was also demonstrating his intimate knowledge of temple history and showing off his martial arts prowess.

Such spectacular acts, by making mythohistory tangible and visible, are critical events in the production of both the collective and the individual actors. Here, an aesthetic moment must be understood as a kind of quantum event: the performer improvises on a set of changes known to all the players and listeners; at the same time, the players and listeners become a collective unit in their intersubjective realization of commonality in the very act of providing the ground for the player to improvise upon. In this sense, the spontaneous possession trance of the ritual actor is an improvisation on a set of changes, within a set of parameters comprising the discourse of imaginable possibilities. The actor creates himself by focusing the energy of the collective upon himself, while at the same time producing the collective that, in this moment, depends on the charisma of the actor to reproduce itself as a self-identified collectivity.

A-kiat-a's actions would be unacceptable if seen as willed or voluntary, but if the agent of the gesture is a deity then the action takes on a completely different meaning, as does A-kiat-a's life and person out of character. The act manifests and affirms a shared collective narrative as well as identifying A-kiat-a as an essential element of the drama. By eliciting the collective recognition of a shared narrative, members of both groups experience a moment of definition. A-kiat-a's spontaneous act immediately reproduces and defines the two groups along their lines of connection and antagonism. A-kiat-a's own group certainly recognizes the creative style of the actor himself, as conversations with temple insiders have confirmed on many occasions. Yet, at the same time, they must misrecognize the agency of the act; similarly, A-kiat-a cannot but affirm (and I have no reason to think that he doesn't believe with complete conviction) that his outbursts are the intentional acts of his patron deity working through him as a possessed medium.

A-kiat-a lives, in some measure, through his identification with this persona that comes to be through creative, communicative acts within the temple milieu. The temple group, in turn, comes to be through a process of inversion whereby the individuals that comprise the group produce the group through the labor of their practice; practice that, of course, emerges from participation in and is limited by the expectations of the collectivity.

Conclusion

Paradoxically, identity always takes multiple referents. Depending on context, the salient "collectivity" with which one identifies may refer to anything from a small group of sworn brothers to a temple community, locality, Taiwan, and so on. The larger, extended collectivities do not simply encompass the smaller, local ones, for each level of collectivity is dialectically linked with both higher and lower levels. Identifying oneself as part of a temple community in Taitung entails simultaneous identification with, say, a faction or clique within the temple as well as the temple's place in the local Taitung ritual hierarchy, and the political networks of city and county; but also and at the same time a set of extended communities[28] that may include the ideological community of Taiwanese, the community of consumers of global styles (and too many others to enumerate here). In practice, these "logical levels" of identity are not discrete entities at all: while individuals identify (in discourse and practice) with different social collectivities, personal attributes, and collective representations and at different moments, each act of identification (whether communicating to oneself or others) is premised upon the mutual creation of the included and the excluded.

Following Lacan, identification entails a division of the self, "the constitution of the subject in the field of the Other" (Lacan 1977: 208). The subject emerges "in the field of the Other," be that the mother in the earliest stages of development or the "alienating symbolic network" (Žižek 1989: 46) of social relations through which we impose on ourselves the impersonal expectations of the social order. In one sense, this dialectic suggests the subordination of the self to the collectivity (Žižek 1989: 113): Whether "keeping one's place" or actively strategizing to accrue power and prestige, individuals—collectively and nondiscursively—make real the collectivities that they themselves rely on to define the limits of their world (Bourdieu 1977, especially 164–172). At the same time, not only is alienation in this sense a prerequisite to social production, it is a necessary precondition for the generation of desire to produce oneself as a subject in "the field of the Other," in other words, to acquire an identity.

For A-kiat-a, Phik-kut-a, and A-peh, self-presentation in ritual performance is just such an act of identification. Within their separate but overlapping social networks, participation in temple processions constitutes a serious expression of personal conviction and solidarity with the collectivity. There is certainly, then, a degree of continuity of meaning for temple processions that transcends their relocation from local to translocal and even transnational events. Nevertheless, for A-peh and Phik-kut-a, and to a different and perhaps lesser extent for A-kiat-a as well, the legitimation afforded by official sponsorship and notoriety of popular appeal have already become constitutive elements of their identities as procession performers.

The ritual roles they have chosen and their styles of performance, however, are distinct. As even such brief and fragmentary accounts attest, for these three men at least, trance performances and testimonials are linked to unique elements of their personalities and social lives. Differences notwithstanding, all three cases illustrate, in much the same way the ideological nature of identification. Relying again on Lacan's terminology, we can discern in these performances the dialectical play of "imaginary identification" and "symbolic identification," "image" and "gaze," recapitulating the split subject and thereby (as a kind of ritual production) objectifying the very process of identification.

It can also be argued that the aggressivity and competitive passion (including both self-directed and other-directed violence) that so dramatically characterize temple processions and trance performances are, in part, controlled and sanctioned manifestations of libidinal frustration. Contained within the limits of ritual, the primal rage and frustration otherwise repressed is released in the trancer's frenzy. I would also argue that the violent energy of such trances is implicitly (and in some ways, ritually acknowledged to be) an expression of infantile sexuality. A regression to infanthood is implied in the inarticulate outbursts, foaming at the mouth (drooling), and self-mutilation; and acknowledged in the fact that entranced *tangki* are simultaneously fawned over and feared by the assistants, not to mention their public nakedness (and donning of the "diviner's bib"). Indeed, all ritual trancers are called *tangki*, "divining child," and these "children" are almost universally aggressive and unpredictably tempermental—but only within the frame of the ritual.[29]

As Žižek explains "imaginary identification is identification with the image . . . representing 'what we would like to be'" (Žižek 1989: 105).

For A-kiat-a this is the identification with the popular image of the capricious but righteous warrior. For A-peh and Pik-khut-a, this ritual role further extends their social personae, which are also modeled on the public fantasy of the Chinese knight-errant. Playing these roles is an active, creative playing-out of a personal and collective fantasy, an act of self-constitution. Yet, this act draws an ideological veil over the concealed "symbolic identification" that supercedes the imaginary: the experience of creative autonomy "is for the subject the way to misrecognize his radical dependence on the big Other" (Žižek 1989: 104). In other words, the performer–subject enacts his own subordination to the collectivity, even as the collectivity is produced through his creative labor. Only through this identification can the performer's underlying desire to satisfy the expectations of his alienated Other be gratified. Here we can see that identity produced through imaginary identification (i.e., identity which can be represented, asserted, denied, debated, negotiated, won, or lost) is a deception; or, as Žižek puts it, "imaginary identification is always *identification on behalf of a certain gaze in the Other*" (Žižek 1989: 106; italics in the original).

Trance performance enables individual subjects to produce themselves as cultural subjects in an act of imaginary identification. Yet in realizing and valorizing the autonomous subject (epitomized here by the mythohistorical knight-errant), the performers are reproducing the moral and institutional structures that bind them to the Other—the doxic "I" coterminous with the ideological limit of collectivity.[30]

Notes

1. This paper draws on field research in Taiwan supported by grants from the Fulbright Foundation, American Council of Learned Societies, the China Times Cultural Foundation, and the Pacific Cultural Foundation; Faculty Research Grants from Hobart and William Smith Colleges; and a College of Liberal Arts Vision Grant at the University of Texas/Austin.
2. Official publications through the early 1980s are littered with articles condemning what the official KMT line held to be the "excesses" of popular religion, suggesting methods of control and even outright suppression (for examples of the more extreme position, see Luo Yunjia "Let us Utterly Eliminate Superstition" [chedi saochu mixin] in Police Magazine [*Jingcha Zazhi*] 5: 82, 1959; and Fan Guoguang, "How can we eradicate superstition?" [ruhe pochu mixin] in Taiwan Provincial Government (*Taiwan Shengzheng*) 3: 29–30). Few, if any of these pieces were written by highly-regarded scholars, however, although the party's control of academic institutions and public discourse prevented open advocacy and led most scholars,

anthropologists among them, to carefully circumscribe ideologically sensitive issues and avoid any suggestion of advocacy.

3. Throughout this period, the Council for Cultural Affairs (Wenhua Jianshe Weiyüanhui) has played a key role in supporting (and thereby, to some extent directing) the practice and preservation of "native culture" in Taiwan. While the mandate of the Council touches on just about every conceivable definition of "culture" (an English version of the Council's principles, guidelines, and current activities can be found at http://www.cca.gov.tw/intro/index_e.html), Council leadership has generally managed to get around ideological pressure to privilege elite Chinese and Western arts at the expense of local practice. This is possibly due to the influence of the Council's first chair, the highly respected and influential anthropologist Chen Chi-lu (Chen Qilu).
4. Notable among these was the popular "TV news magazine" "Ninety Minutes." Designed to appeal to an increasingly cosmopolitan, urban, middle-class audience, this program ran several features on popular religion in the late 1980s and early 1990s.
5. See, for instance, Kataoka 1981 (1921) and Suzuki 1978 (1934).
6. This, of course, would seem to entail a further contradition, since the *jiang* were transformed through defeat by a deity. Often the *jiang* serve a group of deities, only one of whom is truly able to keep them in line, as is the case with the Eight Military Retainers who serve the Emperors of the Five Blessings (Wu Fu Da Di). There are many different mythohistorical origin stories of the Military Retainers, but the following, related to me by a ritual master whose father was from Fuzhou is fairly typical: According to this account, the Military Retainers were originally a pack of malicious mountain spirits (M. *shan yao*) who lived in a cave on Mt. Lu, and were therefore known as the "Eight Demons of Lushan" (Lushan Bayao). These demons were so malicious and powerful that the orthodox deities, including the Jade Emperor, were unable to control them. At this point Liu Wenda, third of five plague deities identified as either the Ministers of Plague or the "Five Lords Who Manifest Efficacious Power" (Wu Xianling Gong) stepped forward. An expert in martial and Taoist arts and the third of the Five of the Five Blessings (Wufu Dadi) only he had sufficient skill to challenge the mountain spirits. He went to Mt. Lu and promptly subdued (*shouhu*) the demons. After defeating them, he held them captive rather than destroying them. To repay the favor of sparing their lives, the demons swore eternal allegiance to Liubu and vowed to help him "save the world." The five plague deities then presented the demons of Mt. Lu to the Jade Emperor. The Jade Emperor, deeply impressed with (and, according to my informant, fearful of) Liubu's power "promoted" the five to the status of "Great Emperors of the Five Blessings" (Wufu Dadi) and put Liu Wenda, now known as Liubu Dadi (Great Emperor of the Liu Division) directly in charge of the demons. Thenceforth the reformed demons became the military retainers of the Great Emperors.

   The incantation used to invite the spirits of the Military Retainers at the Dongxing Temple in Dali, Taichung, gives a slightly different version.

Here, the Military Retainers were originally the ghosts of bandits in the area of Mt. Song (Songshan, Siong-san):

> Respectfully [I, the ritual master] invite the Ten Military Retainers [who are] Great Deities:
>
> Their penetrating spirit-power is vast [so that they may] save the common people.
>
> Originally generals of Mt. Song, they swore fealty to my master [Liubu Dadi], who employs them as his vanguard.
>
> My master has been sent to investigate good and evil, to absorb the Five Poisons in Taiwan Prefecture according to the command of the Jade Emperor.
>
> The Duke of Awesome Subduing [Power who] Manifests Efficacy appears, he exorcises evil and cures illness, extending his awesome supernatural efficacy.
>
> He has bestowed on the Ten Generals the privilege of carrying the talismans [denoting their official appointment]; my master arrays his myriad troops and horses.
>
> Leibu [the "captain" of the Military Retainers, represented in processions as "Bailiff"], my master, has the efficacious power to expel evil; when he unfurls the magic talisman *ghosts and gods* alike are terrified. [emphasis added]

7. The Zhiben Valley, for instance, was first developed as a hot springs resort by the Japanese during the colonial period, and until the 1990s the local Puyuma people were able to enjoy the hot spring water in pits dug in the rocky riverbed. During the mid- and late 1990s, the once-quiet valley, was built up to nearly urban levels and the hot springs water commandeered by a few of the larger hotels, leaving little for downstream establishments and nearly nothing for the natural riverbed swimming holes. The butterflies that appeared every spring in a profusion of varieties and numbers have all but disappeared over the short span of twenty years. For an overview of the challenge to preserve the natural environment, see Su Chengtian, "Dongbu Jingguan Weihu Keti Zhi Tantao" (Examining Projects to Preserve and Maintain the Scenery of the East Coast) (1997: United Daily News Cultural Foundation). The rationale and outline of parts of the development plan can be found in the 1983 Taitung County Government document entitled "Taitung xian zhengti fazhan guanguang gangyao jihua baogao shu" (Outline Plan Report on the Integrated Development of Tourism in Taitung County).
8. Once for most households, thrice for temples or for the altars of important temple members, politicians, etc.
9. Temple troupes perform this ritual at every temple they pass, but usually a visit to a private altar entails an invitation from the household or implies a social or political connection between the household and the temple.

The homes and businesses of all temple members, for instance, will be visited without explicit invitation.

10. Donald Sutton has examined such performance troops in his recent articles such as Sutton 1990 and Sutton 1996.
11. Sutton (1996 and elsewhere) argues for a more nuanced categorization based on the dramatic content of the respective performance troupes. "Civil" and "Martial" are common emic categories, however, and suffice for the purposes of this paper.
12. There are a number of extant translations of the name, including "Infernal Generals." I feel, however, that "Military Retainers" more accurately describes their historical role and is a more literal translation. The Military Retainers belong to the same family of martial procession troupe as the lion dancers, the Song Jiang Battalions and the martial "mannequins" (H. *toa-sin ang-a*) described by De Groot (1982), Sutton (1996), Suzuki (52ff.), and others. There is some justification for the speculation, however, that the Military Retainers are a transformation of another set of popular (but historically obscure) deities, including the so-called "Eight Spirits" (M. *bashen*) or "Eight Generals" (*bajiang*).
13. This directive is written into the Heavenly Decree which is read aloud and burned, the ashes mixed with water and shared, in turn, by the Retainers before they can fall in and begin their procession march.
14. Some informants actually refer to Military Retainers as *tangki*, while others argue that there is a categorical distinction owing, in part, to very different modes of training and function as well as the difference noted here, namely, that *tangki* can only be identified with their patron deity when actually in trance, whereas Military Retainers, in trance or not, are in a constant state of semi-possession. Military Retainers, moreover, derive their efficacy from working as a unit, whereas possessed tangki perform an individual role, either in processions or at altar-side healing/oracle sessions.
15. In the past, troupe members were usually selected through divination. When communities were smaller and mobility more limited than it is today, it was far more feasible to organize a troupe in this way. With the combination of compulsory military service and social and economic pressures limiting the time young men can devote to traditional collective activities, troupe members are now often recruited without the formality of divination.
16. This provides the actor with some measure of protection against unclean things that might want to cause him harm. Having once been a Retainer, as the stock phrase has it, "Big tragedies are reduced to minor incidents, and minor incidents are avoided altogether" (*da shi hua wei xiao shi, xiao shi hua wei wu shi*). And as with *tangki*, there is an unspoken assumption that those with a predilection to contact with the unseen realm have "one foot in the grave," as it were (those prone to possession trance are supposed to have lighter "eight characters" [*bazi*] than most mortals). Performing as a Retainer, however, adds time to one's parents' life span (*tian shou*), justifying it therefore as a filial act.
17. There doesn't seem to be a totally satisfactory term to describe what I (and others) have variously called the production of cultural subjectivity, formation of identity, and so on. Here I am using the verb "to identify" in an

unconventional sense that tries to encompass all these meanings—that is, the conscious intention of declaring one's affiliation with a group, idea, tradition, etc.; the non-deliberate but "conscious" aspects of the phenomenology of this process; and the intersubjective processes through which individuals and groups are produced and reproduced as discrete, "identifiable" entities.

18. Cf. Sutton 1996, especially 218–221. Sutton distinguishes three types of Military Retainers Troupe, namely "traditional," "commercial," and "self-mortifying" (219). As Sutton himself notes, the three types are not sharply distinguished from each other. The distinction is relative and generally quantifiable: even staunchly "traditional" troupes will accept ritual offerings of cash to offset expenses and host a banquet for troupe and temple members, for instance, although unlike "commercial" troupes the cash is not distributed to the participants, who are nominally volunteers. All troupes have members who are adept at trance, and most "commercial" troupes feature self-mortification, although the generals may be forbidden from wielding bladed weapons and needles and the mortifiers may comprise a separate group.
19. All personal names given here are pseudonyms.
20. "Malu zu," literally "road team." This tongue-in-cheek term refers to the group of young toughs whose main role is to enter (or feign) trance and perform spectacular acts of self-mortification during temple processions. This particular group is well-known for its "performances."
21. See, for instance Sangren 1984; Duara 1988; Chen 1987; Ownby 1996; and Murray 1994. The theme of loyalty and revenge among sworn brothers is also a dominant theme in both the martial arts and gangster genres of popular novels and films in Hong Kong and Taiwan.
22. Which, it should be noted, also entail a closeness and mutual trust that the American variety of machismo precludes, something which often startles and sometimes earns the admiration of non-Chinese.
23. This particular connection is one of "*renqing*," that is, social obligation entailed by friendship, rather than a structured lineage relation between temples. One of the leading temple committee members at Gongrong Temple is the elder brother of Pik-khut-a's closest associate in Taichung. All of them grew up together in Taitung, and trace the beginning of religious knowledge and practice to Zhonghe Temple.
24. Crossroads are, of course, the sites of frequent accidental deaths, and are thus considered places of pollution and danger. Wang Ye and the Military Retainers are specifically agents of exorcism, and spirit mediums frequently become possessed at such places in order to "transact the affairs [of heaven]" (*ban shi*)—that is, capture and eradicate the ghosts of accident victims which may populate the area.
25. Cf. De Groot, *The Religious System of China*, esp. Book IV, *The War Against the Spectres* for an account of similar practices in Xiamen during the last century.
26. That is, the tragic story of the sworn brothers Xie and Fan, whose souls, as Seventh Master and Eighth Master serve the God of Walls and Moats,

as well as being integral members of Wufu Dadi's retinue of Military Retainers.

27. Incense is offered to any seen or unseen manifestation of supernatural power, from the wrongfully injured ghosts at accident sites to temple images to possessed tangki. The offering of incense to Military Retainer actors, whether visibly possessed or not is a sign that (during the procession) they metonymically embody the unseen exorcistic power. The Retainers are thus "possessed" primarily through the power of ritual magic. This is unusual in the sense that this sort of ritual magic is used by *fashi* to manipulate supernatural power, even to channel it through their bodies, but never do they become embodiments of deities per se. On the other side, *tangki* are only seen to embody deities or ghosts when clearly manifesting the signs of possession trance. The Military Retainers appropriate the conventions of scripted, costumed traditional opera, but because they also act spontaneously, they take aesthetic representation one step further, to the complete coextension of sign and referent.
28. "Extended communities" here might also be called "imagined communities" in Ben Andersen's sense, but instantaneous, interactive electronic and digital media have already expanded the community well beyond "nation." What's more, the print-dominated mass media of the pre-electronic age still allowed for the ambiguity of imagination: there is plenty of space and time between a reader and the printed page and between participants in a print-generated discourse. As so many critics have suggested, electronic media are far more efficient mechanisms of control, and are perhaps beginning to take the place of the imagination itself.
29. I explore this line of thought further in another study, in progress at the time of this writing.
30. Contrary to what a common-sense reading of this terminology may suggest, I am not suggesting here that subjects are deceived and unwittingly or unwillingly bound or constrained, but that alienation is intrinsic to the process of ritual (as well as all other modes of social production). As Sangren puts it, "ideological alienation facilitates alienation in the classical Marxist sense of separating the producer from his or her activities or labor . . . Alienation, in other words is not merely a misrecognition or mystification of the productive process, but is also itself productive of the forms and processes of social life" (2000: 2).

References

Anderson, Benedict (1983) *Imagined Communities*. London: Verso.

Bateson, Gregory (1972) *Steps to an Ecology of Mind*. New York: Ballantine.

Bodde, Derk (1975) *Festivals in Classical China*. Princeton: Princeton University Press.

Boretz, Avron A. (1996) *Martial Gods and Magic Swords: The Ritual Production of Manhood in Taiwanese Popular Religion* (Ph.D. dissertation).

Boretz, Avron A. (Bai Anrui) (2001) "Minsu Quyi" (Local Customs) in Taidong Xianshi: Hanzu Pian (History of Taitung County: The Han Chinese): 116–161. Taitung: Taitung County Government.

Bourdieu, Pierre (1977) *Outline of a Theory of Practice*. Cambridge: Cambridge University Press.

Davidson, James W. (1903) *The Island of Formosa: Historical View from 1430 to 1900*. New York: Macmillan.

Chen, Changfeng (1987) *Zhulianbang: xingshuai shimo (The Rise and Fall of the Bamboo Union Gang)*. Taipei: publisher obscure.

Dean, Kenneth (1993) *Taoist Ritual and Popular Cults of Southeast China*. Princeton: Princeton University Press.

Dean, Kenneth, and Zhenman Zheng (1993) "Group Initiation and Exorcistic Dance in the Xinghua Region" in *Minsu Quyi* 85: 106–214.

DeGlopper, Donald (1995) *Lukang: Commerce and Community in a Chinese City*. Albany: State University of New York Press.

Dong, Fangyuan (1982) "Taiwan minjian zongjiao jiyi—Song Jiang Zhen" (Skill and Art in Taiwanese Popular Religion: Song Jiang Battalion) *Zhong'guo luntan* 13.8: 25–32.

Duara, Prasenjit (1988) *Culture, Power, and the State: Rural North China, 1900–1942*. Stanford: Stanford University Press.

Feuchtwang, Stephan (1992) *The Imperial Metaphor: Popular Religion in China*. London: Routledge.

Elvin, Mark (1985) "Between Earth and Heaven: Conceptions of the Self in China" in M. Carrithers, S. Collins, and S. Lukes (eds.), *The Category of the Person: Anthropology, Philosophy, History*. Cambridge: Cambridge University Press.

Goffman, Erving (1959) *The Presentation of Self in Everyday Life*. Garden City, NY: Doubleday.

Gong, Pengcheng (1987) *Da Xia (The great knight errant)*. Taipei: Jinguan.

Groot, J. J. M. de (1982 [1892–1910]) *The Religious System of China, Its Ancient Forms, Evolution, History, and Present Aspect, Manners, Customs and Social Institutions Connected Therewith*. 6 vols. Taipei: Southern Materials Center.

Huang, Wenbo (2000) *Taiwan Minjian Yi Zhen (Temple Procession Troupes of Taiwan)*. Taipei: Changmin Wenhua/Formosa Folkways.

Huang, Wenbo (1997) *Taiwan Minjian Xinyang yu Yishi (Folk Religion and Ritual in Taiwan)*. Taipei: Changmin Wenhua/Formosa Folkways.

Jordan, David K. (1972) *Gods, Ghosts, and Ancestors: Folk Religion in a Taiwanese Village*. Berkeley: University of California Press.

Kataoka, Iwao (1981 [1921]) *Taiwan fengsu zhi (Record of the customs of Taiwan)*. Taipei: Da Li.

Katz, Paul (1995) *Demon Hordes and Burning Boats: The Cult of Marsha Wen in Late Imperial Chekiang*. Albany: State University of New York Press.

Katz, Paul (Kang Bao) (1997) *Taiwan de Wangye Xinyang (Taiwan's Cult of Wangye)*.

Kleinman, Arthur (1980) *Patients and Healers in the Context of Culture: An Exploration of the Borderland between Anthropology, Medicine, and Psychiatry*. Berkeley: University of California Press.

Lacan, Jacques (1977 [1973]) *The Four Fundamental Concepts of Psychoanalysis*. New York: Norton.

Lacan, Jacques (1977) *Écrits: A Selection*. New York: Norton.

Lambek, Michael (1981) *Human Spirits: A Cultural Account of Trance in Mayotte*. Cambridge: Cambridge University Press.

Lewis, I. M. (1989 [1971]) *Ecstatic Religion: A Study of Shamanism and Spirit Possession*. London: Routledge.

Lin, Meirong (1997) *Zhanghua Qüguan yü Wuguan* (*Ritual Music and Martial Arts Organizations in Chang-hua County*); 2 vols. Chang-hua: Chang-hua County Cultural Center.

Liu, Zhiwan (1983) *Taiwan Minjian Xinyang Lunji* (*Essays on Taiwanese Popular Religion*). Taipei: Lianjing.

Loizos, Peter (1993) *Innovation in Ethnographic Film*. Chicago: University of Chicago Press.

Lu, Mingzhi "Song Jiang Zhen de wenxian yanjiu" (Documentary studies on the Song Jiang Battallion) *Shilian Zazhi* 14 n.p.

Obeyesekere, Gananath (1981) *Medusa's Hair: An Essay on Personal Symbols and Religious Experience*. Chicago: University of Chicago Press.

Ownby, David (1996) *Brotherhoods and Secret Societies in Early and Mid-Qing China: The Formation of a Tradition*. Stanford: Stanford University Press.

Murray, Dian H. and Qin Baoqi (1994) *The Origins of the Tiandihui: The Chinese Triads in Legend and History*. Stanford: Stanford University Press.

Roget, Gilbert (1985) *Music and Trance: A Theory of Relations Between Music and Possession*. Chicago: University of Chicago Press.

Ruhlmann, Robert (1960) "Traditional Heroes in Chinese Popular Fiction," in Arthur F. Wright, ed., *The Confucian Persuasion*, 141–176. Stanford: Stanford University Press.

Sangren, P. Steven (1984) "Traditional Chinese Corporations: Beyond Kinship" in *Journal of Asian Studies* 43.3: 391–415

Sangren, P. Steven (1991) "Dialectecs of Alienation" in *Man* 26: 67–86.

Sangren, P. Steven (2000) *Chinese Sociologics: An Anthropological Account of the Role of Alienation in Social Reproduction*. London: Athlone Press.

Shi, Wanshou (1983) "Jiajiang tuan—tianren heyi de xunbu zuzhi" (Military Retainers troupes—a transcendental police organization) in *Tainan wenhua* 22: 47–65.

Sutton, Donald (1990) "Ritual Drama and Moral Order: Interpreting the Gods' Festival Troupes of Southern Taiwan" in *Journal of Asian Studies* 49.3: 535–554.

Sutton, Donald (1996) "Transmission in Popular Religion: The Jiajiang Festival Troupe of Southern Taiwan," in M. Shahar and R. Weller, eds., *Unruly Gods: Divinity and Society in China*, 212–249. Honolulu: University of Hawaii Press.

Suzuki, Seiichiro (1978 [1934]) *Taiwan jiuguan xisu xinyang (Old customs, practices, and beliefs of Taiwan)*. Taipei: Zhongwen.

Tamkang University, Department of Chinese Literature (1993) *Xia yü Zhong'guo wenhua (The knight errant and Chinese culture)*. Taipei: Xüesheng Shujü.

Turner, Terence (1991) "We are Parrots, Twins are Birds: Play of Tropes as Operational Structure" in James W. Fernandez, ed., *Beyond Metaphor: The Theory of Tropes in Anthropology*, 121–158. Stanford: Stanford University Press.

Turner, Terence (1995) "Representation, Collaboration and Mediation in Contemporary Ethnographic and Indigenous Media" in *Visual Anthropology Review* 11.2: 102–106.

Ward, Barbara E. (1979) "Not Merely Players: Drama, Art, and Ritual in Traditional China" in *Man* 14:18–39.

Weller, Robert (1987) *Unities and Diversities in Chinese Religion*. Seattle: University of Washington Press.

Xilai Temple Committee n.d. *Wufu Dadi: Xilai An Jingli Zhuan (The Great Emperors of the Five Blessings: The History of the Xilai Temple)*. Tainan: Xilai Temple.

Žižek, Slavoj (1989) *The Sublime Object of Ideology*. New York: Verso.

Chapter 9
Anthropology and Identity Politics in Taiwan: The Relevance of Local Religion[1]

P. Steven Sangren

I initially drafted this chapter subsequent to a conference on "Economy and Society in Southeastern China," East Asia Program, Cornell University, 2–3 October 1993. The conference presentations of Murray Rubinstein and Kenneth Dean set me thinking, both about the issues their papers raise regarding the processes of what might be termed "identity construction" and about the ways identity issues have emerged, more broadly in China studies and elsewhere. I found Dean's and Rubinstein's papers particularly interesting, because each focuses on the study of religious cults in Fujian that are counterparts of cults that have been the objects of much of my professional research and writing. Moreover, their papers (see also Dean n.d.a and n.d.b) confirm that local cults continue to play (or, at least, are emerging to play again) a constitutive role in the formation, perpetuation, and transformation of local and regional identities in Fujian.

Subsequent to completing the first draft of this paper, I spent six months in Taiwan as a Fulbright Research Scholar and Visiting Professor at National Tsing Hua University. Serendipitously, I was present during President Li Denghui's visit to Cornell University, and I had the opportunity to observe the lively discussions between various Chinese and Taiwanese groups associated with the event. More recently, I was in Taiwan in August 1995 in the midst of the PRC's escalation of belligerent military exercises explicitly intended to influence Taiwan politics.

I make particular note of these circumstances because of the rapidity of change in the politics of "culture" in Taiwan: Although the major conceptual points of my first version withstand revision in light of recent circumstances, I have substantially reassessed (upward) the degree to which identity issues are driving politics (and the converse) in Taiwan. During the December 1994 elections for governor of Taiwan, and for mayor, in such major cities as Taipei and Kaohsiung, the KMT (Guomindang) struggled (with limited success) to prevent ethnicity from becoming *the* major issue of the campaign, and some DPP (Democratic Progressive Party) candidates struggled just as energetically to affiliate the KMT (increasingly dominated by ethnic Taiwanese) with an increasingly discredited legacy of Mainlander-dominated government.

The recently inaugurated New Party acted as something of a "spoiler." Mainly representing the disaffected young Mainlanders, the New Party attempted to make the election a test of national patriotism, defined in what to many in Taiwan seems an anachronistic reference to KMT pretensions to be the legitimate government of all of China. The New Party may have embarrassed the KMT slightly by exposing the enormous gulf between the KMT's official line with regard to Chinese unification and its practical policies; but in Taipei's mayoral contest, the New Party clearly damaged Huang Dazhou (the incumbent KMT mayor who managed only an embarrassing third-place finish behind the winning DPP candidate Chen Shuibian and second-place NP candidate Zhao Shaokang). However, the New Party's major impact seems mainly to have been to make the KMT seem a moderate alternative between two extremes.

In sum, Fall of 1994 was an exciting time to be in Taiwan. In the relatively brief decade or so since Chiang Ching-kuo began to liberalize Taiwanese politics, Taiwan has developed what one can only characterize understatedly as a "lively" multiparty democracy. These developments have produced both optimism and anxiety, and the many Taiwanese with whom I spoke, from across the spectrum of political commitment, find themselves torn between the two sentiments. More to the point of the present paper, the embeddedness of identity issues in practical life and politics is also abundantly evident in academic life. To cite only the most pertinent example, it is difficult for Taiwanese scholars (or their foreign colleagues) to sustain a stance of impartial objectivity in the study of, for example, my own specialty—"folk religion"—when one is likely to be criticized either for assuming and legitimating a category of "Chinese culture" that diminishes the uniqueness of Taiwanese experience, on the one hand, or of caving in to the fashionably and politically correct emphasis on Taiwanese specificity, on the other.

Introduction: The "Taiwan Question"

There are striking similarities in the ways religious activities demarcate social solidarities in Fujian to processes familiar to us from Taiwan-based ethnography (Feuchtwang 1974; Harrell 1982; Weller 1987; DeGlopper 1974; Schipper 1977; Sangren 1987; Baity 1975). Dean's work suggests that there may also be some significant differences (for example, in the relative weighting of Taoist "liturgy" in local religion in Fujian).[2] Obviously, the political and economic circumstances framing emergent local and regional identities in Fujian differ significantly from those that have constituted the wider social context in Taiwan. The specification and interpretation of such differences constitutes an exciting new arena for specialists, and perhaps has a potential contribution to make to practical dialogue in the nexus of Taiwan, *Min'nan*, and Chinese politics as well.[3] The question I address here is, in short: In the rapidly changing economic and political circumstances facing Taiwan and Fujian, how is religiously motivated collective action implicated in the reproduction of longstanding lines of demarcation of social solidarity, and (equally significantly) also employed in the production of new definitions of community and identity?

Questions of "identity"—particularly whether or not present-day Taiwanese constitute an identity that ought not be considered "Chinese"—are hotly contested in Taiwan. Liberalizations of the past decade or so have made such disagreements much more public than was the case during my initial fieldwork in the mid-1970s. In Taiwan, both explicitly and implicitly, differences of opinion and commitment regarding Taiwanese and Chinese identity inform a broad spectrum of public discourse—in the arts, in the academic world, in educational policy, and in mass culture—as well as in politics as such.

Opinion ranges from Taiwanese who fervently resist the notion that they are Chinese to old-guard Mainlanders whose nationalist aspirations for a unified China under KMT rule remain undiminished. The sentiments of most of the island's residents, the Mainlanders and the Taiwanese, fall between these poles. Although an argument can be made that a general trend toward moderation is perceptible during the past decade or so (Bosco 1992a), the issues continue to animate Taiwan's political scene. The mainland government is obviously also an interested party and felt presence—increasingly since its recent military posturings. Its official policy is to insist on Taiwan's eventual political reunion with China; the ever-present possibility that it might resort to force to implement this goal deeply informs Taiwan politics. Moreover, the

PRC's heavy-handed bullying in the arena of international relations[4] seems, for the most part, to unite otherwise fractious political opinion.

The political salience of the "Taiwan question," however, should not divert our attention from the fact that disputes over identity issues are not limited to the longstanding tensions between Taiwanese and "Mainlanders," or between the PRC and Taiwan. Most obviously, the identity status of Taiwan's aboriginal peoples (of Austronesian heritage) further complicates any straightforward definition of what constitutes "Taiwaneseness";[5] more demographically significant are distinctions between the Taiwanese who trace their ancestry to Hakka areas of China's Southeast Coast macroregion, and to Zhangzhou and Quanzhou prefectures.[6]

To such subethnic distinctions, one must add those distinctions based on a nested hierarchy of local communal identities expressed in territorial cults. Local identities are represented and contested in collective ritual action at several levels (neighborhoods, villages, towns and their hinterlands, and city-centered local systems).[7] Although most performances of collective ritual entail some elements of community boundary marking, symbolic references to wider, hierarchically encompassing communities or cultural spheres are, I have argued, necessarily invoked (Sangren 1993). Moreover, many communal rituals manifest, to employ currently ascendant jargon, "contested" interpretations of how local communities ought to be conceived and of the nature of relations among differently constituted communities.[8]

In sum, local religion and collective ritual practices constitute an important arena of production of collective identities. Included in these identities are not only such obvious contrasts as the Taiwanese versus mainland identity, or even of Zhangzhou versus Quanzhou heritage, but also much more locally specific, frequently crosscutting allegiances. It is not the case in Taiwan that "all politics is local," but lines of local solidarity mediate the impact of wider identity issues.[9]

Hot on the heels of political liberalization in Taiwan has come the liberalization of travel restrictions. Since 1987, the Taiwanese have been traveling in large numbers to the mainland, particularly to Fujian.[10] The number of Taiwanese visitors to China has grown from some 400,000 in 1988 to more than 1,500,000 in 1993.[11] As both Dean and Rubinstein document (see also Sangren 1988b, 1993), much of this travel is religiously motivated. The close linkages between collective religious action and various notions of identity, in part, inspires this travel. I shall suggest here that such travel has the potential to modify notions of identity in both travelers and, more subtly, their hosts.

The entire situation is so complex and in such a flux, that even if I had extensive fieldwork-based interview data at my disposal,[12] it would be ill-advised to attempt to produce general conclusions or predictions regarding the shape of emerging identities in Taiwan or the Southeast Coast. However, I think that it is possible to produce a preliminary map of the field possibilities and to describe the range of contestants.

An additional reason for caution here ought to be obvious, but should nonetheless be stated explicitly: Identity politics is inherently a very sensitive topic. Several American anthropologists have been the objects of scathing published criticisms;[13] some such criticisms are motivated by the passionate commitments of their authors to particular understandings of where the lines of legitimate cultural (and political) identity ought to be drawn.

I hope that readers with justifiable and deeply felt commitments regarding such issues will be indulgent and allow me to evade, to some degree, judgments about the logic of competing substantive claims. It has become a cliche in anthropology to observe that all identities are constructed; in other words, who and what we are is to a considerable extent, the product of our own "fictional"[14]—that is, collectively *produced*—discourses. Individually and collectively, we are who we define or imagine ourselves to be—subject, of course, to others' powers to accept or to limit such self-constructions. Identity, in short, is a product of individual and collective processes of construction—that is, it is a product of social and individual histories—and encompasses socially relevant categories ranging from gender and profession to ethnicity and citizenship.

To assert such truisms, however, is not to argue that constructed identities are not real; they are real to the extent that they define individual and collective senses of identity and to the degree that such commitments motivate individual and collective *action* or (to use the fashionable term) *practices*.

For example, with regard to the argument that Taiwanese culture and history constitute evidence for a collective identity superseding an overarching Chinese one, anthropology (particularly as practiced in the writings of Western anthropologists) ought not to make hasty judgments.[15] If, in the course of presently unfolding developments in Taiwan, such arguments hold sway (in other words, if such constructed identities motivate effective collective action), then with typical disciplinary perspicacity (that obtained by hindsight), anthropologists (like historians) can link the production of emergent Taiwanese identity to the historical and cultural circumstances that produced it. Conversely, if a competing regional focus on, for example, the *Min'nan* area emerges as a more potent locus of collective action, then our explanatory narratives will follow suit.

The foregoing observations boil down, in part, to the obvious fact that it is the Chinese, the Taiwanese, and *Min'nan* people themselves who will determine the parameters and contextual significance of their variously defined and practiced identities; it is not for Western scholars to pass judgment upon which among the various possibilities has the greatest historical, cultural, "objective," warrant. It would also be unwise to prognosticate; predicting the future lines of social solidarities could easily be interpreted as advocacy.

Let me add, however, that in my view anthropology can shed some light on identity construction as a sociocultural process—perhaps most significantly by making explicit the ways in which history and cultural identity are employed in situationally specific "rhetorics of legitimacy"[16]—rhetorics that become part of emergent social realities.[17] Discussions like this one can thus record some of the possible fault lines, some of the issues that constitute the current context within which identity politics is conducted.

Tradition

In the absence of systematically collected interview data, I would have to proceed anecdotally. In August 1992, I made my first visit to Fujian to visit the home temple of the Mazu cult on the island of Meizhou in Putian County. Returning to Xiamen, I decided to stopover in Quanzhou, a city that figures prominently in narratives of Taiwanese history and identity. Of course, I visited the famous temples and pagodas of the city, but I also made it a point to seek out a shop where I had heard where Chinese hand-puppets could be purchased.

My interest in these puppets is, in part, inspired by my interest in the mythical characters that some of them portray. No doubt as a consequence of my research, my favorites among them are gods, demons, and immortals rather than courtesans, warriors, and officials. In this respect, my desire to acquire these particular objects is somewhat idiosyncratic, but my motives were not dissimilar in principle from those of many Taiwanese travelers who, since 1987, have been visiting Fujian in large numbers.

My thesis, in brief, is that many Taiwanese who travel to Fujian are inspired by a notion of Fujian as the ancient and authentic source of their own cultural traditions, their identity. Fujian locales figure prominently in the written and oral tradition of the Taiwanese; since the late 1980s, the Taiwanese have for the first time since World War II been able to visit some of these places. Although I am not Taiwanese, I spent several years in Taiwan studying, specifically, local religion and history, and my attitudes toward Fujian as a source of "tradition," both Chinese and more specifically Taiwanese, bear the mark of this experience.

The cultural linkage between Taiwan and Fujian has obvious objective historical bases. As stated earlier, the largest percentage of Taiwan's present-day population are native speakers of *Min'nan hua* ("Southern *min*") who trace their ancestry to immigrants from the Zhang-Quan region. The settlement history of Taiwan preserved the significance of distinctions of place of Fujian origin down to the prefectural and sometimes even county and village level (Sangren 1987; Knapp 1976).[18] For example, in Daxi *zhen*, where I conducted my dissertation fieldwork (1975–77), about 90 percent of the township's population quite explicitly identified themselves as Zhangzhou *ren* (Zhangzhou people); in Sanxia, the township bordering Daxi to the north, about 90 percent hail from Anxi county in Quanzhou *fu*; in Longtan, to the southwest, most were Hakka speakers from southern Fujian and northern Guangdong.[19]

The clustering of people along such lines is the result of complex historical circumstances that differ in different locales; among the most important were that (1) in many areas settlement was organized by large landlords (*dazu*) who recruited tenants from their own native places and (2) Taiwanese settlement history was characterized by violent competition for control of land and water in which distinctions based on place of Fujian origin formed the lines of opposition in factional fighting (Lamley 1981). These historical circumstances find present-day social significance in the role of local territorial cults. Community membership in Taiwan is defined with reference to, among other criteria, territorial cults. Many of the territorial-cult deities that inhabit village and town temples are, in turn, associated with specific places in Fujian.

For example, the most prominent deity in local territorial cults in Daxi is Kaizhang Shengwang. The cult's traditional "ancestral temple" (*zu miao*) is in Zhangpu, Fujian; and its focal symbol is a deified historical personage associated with the pacification of the Zhangzhou region (Dean 1993a). Qingshui Zushi, a god associated with Anxi county in Fujian, heads the Sanxia-centered local system, and temples to Sanguan Dadi are popular in Hakka areas.

The significance of localities in Fujian for Taiwanese conceptions of local identity is thus not limited to historical connection alone, but also persists in the narratives that constitute local identities and in the distinctions among them in present-day Taiwan itself. In Taiwan, to identify oneself as a Zhangzhou *ren* or as a Quanzhou *ren* thus conveys both the sense that one's ancestors hail from Zhangzhou or Quanzhou and that one belongs to the category of those Taiwanese who trace their ancestry to those locales.[20]

This subtle point might be clarified if we note that the distributions of people who trace their ancestry to divergent mainland locales, although clustered, are not contiguous. For example, Yilan county in northeast Taiwan is a Zhangzhou dominated area, and there are significant Hakka concentrations in both Pingdong and Xinzhu counties. The result is that terms like Zhangzhou *ren* and Quanzhou *ren* have taken on rather odd, almost totemic qualities in Taiwan. No longer directly territorial designations of local cultural identity, they now also convey the sense of different types of people. The appellation "subethnic" thus glosses a nexus of historical and cultural factors when applied to such categories. Yet all of these groups are "Taiwanese"—in the commonly employed sense that their forebears settled in Taiwan prior to 1896 when the island came under Japanese imperial control—as opposed to "Mainlanders" (*waisheng ren*) who arrived after the island's retrocession to the Republic of China in 1946.

The nexus of associations linking Taiwanese local distinctions to locales in Fujian is most directly expressed in the resurgence, since 1987, of religious pilgrimages to sacred sites and temples. Dean and Rubinstein summarize these circumstances well (see also Sangren 1988b, 1991, 1993). Less explicitly, however, a similar pursuit of what I would term "objectifications" of identity seems to motivate even tourists who travel to Fujian for other than overtly religious purposes.

To return to the puppet shop, while I was pondering what "objectification" of Chinese "traditional culture" to purchase, I became involved in a conversation between the manager of the shop and another customer. The other customer was the proprietor of a *fojyu dian* ("religious articles") shop in Tainan, Taiwan. He and his wife were on their fifth visit to Fujian. He told me that he comes to the Chinese mainland mainly to "buy things"—things that possess what he termed "*wenhua jiazhi*" ("cultural value"). He emphasized that he has no "religious" interest in such objects.[21] In short, like me, he is a collector of objects that he deems possess some of the authentic charisma of Chinese or Fujian culture and tradition (*chuantong*).

He also told me that he already possesses a large collection of puppets, images of gods, and other artifacts. While in the store, he seemed most interested in an elaborately embroidered Taoist ceremonial robe. Although he did not buy it, he told the manager that he might return later for it.[22] The fact that such objectifications of authentic traditional culture are now produced for Taiwanese and Southeast Asian overseas Chinese consumption constitutes a paradox of which the customers often seem unaware or unconcerned. Usually overriding any concern over this obvious commodification of tradition and authenticity is the desire of the consumers to purchase (at bargain prices!) their representations. In the case of the Taoist robe, for example, it is consumer demand that has stimulated production.

Another anecdote sheds a slightly different light on Fujian's purchase on Taiwanese identity (and the attempt of Taiwanese tourists to purchase it back?). In the Summer of 1984 (several years prior to the unanticipated relaxation of travel restrictions to China), I interviewed a carver of deities in Lugang. Lugang, once a thriving port, played a prominent role in the settlement of Taiwan, and prides itself on the preservation of aspects of traditional culture—arts and crafts, images of deities, music (DeGlopper 1974). It is, consequently, especially noteworthy that the carver emphasized his Quanzhou origins: His name card displayed the town and county in Quanzhou *fu* whence he traces his lineage.[23]

During my visit, I witnessed the delivery of a modest, made-to-order image of Guangong to some businessmen from the distant city of Jilong. These businessmen were obviously quite wealthy; they arrived in a Mercedes-Benz. The carver's pitch to this group was extraordinary. He claimed that only he, in all of Taiwan, possessed the depth of knowledge of Daoistic principles (*yin* and *yang*) necessary to carve images with proportions corresponding accurately to geomantic conditions (*fengshui*; *dili*). Moreover, he continued, only such images could be expected to achieve the full measure of efficacy of habitation by the god's spirit subsequent to their magical ordination (via a *kaiguang* or "eye opening" ceremony). In other words, the carver's pitch was keyed to a claim to possess exclusive esoteric knowledge linked specifically to a Fujian locale, and this esoteric, authentic "tradition" was claimed to produce images of gods that would be genuinely or especially efficacious (*ling*).

These claims had apparently convinced the customers; the price they paid amounted to at least ten times that generally charged for what appeared to me to be indistinguishable images.

Tradition Unsettled

One wonders what accommodations the carver's pitch must make to current circumstances. Although he can still lay claim to his carver's lineage, the possibility of travel to Fujian must have diminished his claim to be an exclusive source of traditional, authentic (and in this case) magically powerful knowledge.

The opening of China to travel from Taiwan has similarly clouded the claims to cult status of some Taiwanese temples. I have written elsewhere in some detail about this situation (Sangren 1988b, 1993). Here I will add that images of gods recently obtained from the famous Mazu temple in Meizhou, Putian County—conventionally recognized as the temple with the highest status in the Mazu cult, a cult closely associated with the history of Taiwan's settlement from Fujian—have been used by some temples in competition for cult status in Taiwan.

The most prominent Mazu temple in Taiwan, the Qiaotian *Gong* in Beigang, bases its claim to status mainly on its early establishment as a branch of the home temple on Meizhou island in Putian county, Fujian. Subsequently, the Qiaotian *Gong* became the home temple for other Mazu images and temples throughout Taiwan. These subbranches participate in annual pilgrimages of massive proportions (some 4 or 5 million pilgrims visit Beikang every Spring) to restore their images' magical power (*ling*).

Now that travel to Meizhou is possible, however, even Beigang's status has been challenged by some. For example, prior to 1987 the Mazu temple in Dajia, a city in north central Taiwan, sponsored a highly publicized annual pilgrimage on foot to Peikang. However, in 1987, ambitious temple committee members decided to bypass Peikang and proceed directly to Meizhou.[24] Images obtained there were invoked to support a claim to cult status at least equal to that of Beigang.

Dajia's preemptive attempt to usurp status—to create a new tradition (its own status) in the name of a transcendent tradition (Meizhou's)—has, however, not been as successful as its advocates hoped. In part, this lack of success is due to the fact that, although on the one hand the local objectification of the source of tradition (and Mazu's charisma), Meizhou, is now no longer exclusively accessible to claimants like the Beigang temple and the Lugang image carver, on the other hand, the same accessibility of a once unimpeachable authority (unimpeachable, in part, because inaccessible) now subjects claimants to this authority—real world things, places, and people in Fujian—to direct scrutiny.

Take, for example, the images of Mazu obtained by the Dajia temple from Meizhou. To some Taiwanese observers, these mainland-produced Mazu images (at least those obtained soon after the rekindling of pilgrimage visits to Meizhou) appeared overly youthful, attractive. One of my Taiwanese informants joked that they more closely resembled a courtesan than a goddess. This circumstance is unsettling to the

Taiwanese who associate Mazu with a more austere "tradition"; Taiwan's Mazu images are matronly, sedate, dignified, old.

The wry observation of my Taiwanese respondent regarding the cheap, even coquettish, appearance of mainland images raises an interesting issue. On the one hand, in the first flush of "mainland fever" in Taiwan, many Taiwanese have rushed to take advantage of opportunities to leapfrog local claimants to possession of traditional knowledge and its artifacts by seeking higher-statused sources in Fujian.[25] On the other hand, the opportunity to do so has, in some cases, taken some of the luster off the formerly inaccessible authentic roots by revealing their mundane qualities. The authority and legitimacy of formerly transcendent, inaccessible sources of the tradition is now disputable in new and interesting ways—a topic to which I return below.

Tradition Reconstituted

Back in the puppet shop, while the manager was conversing with the Taiwanese customer, his wife asked me if I had ever observed Taoists conduct ceremonies in such robes. When I told her that I that had frequently observed Taoist rituals in Taiwan, she said that in Quanzhou, she had never seen such activities. Dean (1993a) documents the persistence of a vital Taoist tradition in Fujian, and the reemergence of a significant role for it in the production and reproduction of both longstanding and newly constituted lines of social solidarity and community. Yet I believe the sense of loss in the shopkeeper's lament is also noteworthy.

Especially in the cities, the legacy of longstanding official hostility toward such manifestations of "feudal superstition" as local religion and Taoist ritual and magic (particularly during the cultural revolution) has resulted in the denial of opportunities to several generations of Chinese (especially in urban areas) to witness traditional communal rituals,[26] at least in comparison to their counterparts from Taiwan.[27] By the same token, I have heard some Taiwanese academics report having instructed their mainland "informants" in some of the more esoteric details of Daoist ritual.

My hypothesis, in brief, is that just as Taiwanese tourists and pilgrims seek objectifications of an authentic identity in Fujian, in subtle ways, some residents of Fujian perceive in this seeking itself, evidence of persistence in Taiwan of a tradition (or, at least, recognition of its value and legitimacy) denied in the official rhetoric of the People's Republic. Moreover, the fact that the bearers of this avid interest in the value of traditions and objects, representative of an officially derogated "feudal" past, hail from an otherwise relatively modern and wealthy Taiwan (and Southeast Asia) constitutes a paradox obvious to all. One might speculate that the high standing of Chinese tradition in the eyes of such modern, worldly, and wealthy visitors undermines the local credibility in Fujian of official definitions of modernity (i.e., as opposed to tradition), on the one hand, and enhances the legitimacy and appeal of what all might agree is traditional (including, e.g., god worship, puppet shows, Taoist ritual), on the other.[28]

An Emerging *Min'nan* Identity?

I shall now take up an issue anticipated above: One of the unanticipated effects of Taiwanese Mazu pilgrimages to Fujian is that some in Taiwan are now willing to claim cult status for Taiwan temples superior to that of such traditional centers as the temples in Meizhou and in Quanzhou. Interest in acquiring Taiwanese money has frequently been all too apparent in the welcome extended by mainland hosts. Taiwanese pilgrims sometimes express their disappointment in a perceived lack of sincerity in Mazu worship in Fujian; bluntly stated, Mazu worship is a much bigger thing in Taiwan than it is in Fujian (although this situation may be changing). I have heard the argument made (by advocates of the Qiaotian *Gong* in Beigang, for example) that, because the Meizhou temple is new (having been rebuilt with Taiwanese help after it was destroyed during the Cultural Revolution), the most authentic, oldest Mazu tradition—one characterized by an uninterrupted continuity of incense burning (*xianghuo*)—is Taiwan's Beigang Mazu.[29]

Attitudes of partisans of particular temples aside, I do not have a clear quantitative sense of what the majority of Taiwanese devotees of Mazu make of these questions of status priorities among temples. I suspect that the story of the goddess' earthly incarnation associates her irrevocably with Meizhou, and that the history of Taiwanese settlement (and the spread of the cult) will protect the status of temples in Meizhou and elsewhere in Fujian. Taiwanese complaints about the pecuniary motives of mainland temples are unlikely to diminish this status in any fundamental way.

But by the same token, I suspect that the status of Beigang Mazu will also survive attempted usurpations like that of Dajia. The Beigang temple has been consolidating its position atop the Taiwan hierarchy in the Mazu cult for a very long time, and its associations with a Taiwanese sense of history and identity may be indispensable (Sangren 1987, 1988b).

If the number of pilgrims who visit annually can serve as a valid measure of the temple's caché, it seems undiminished by recent developments. Events of the late 1980s may have unsettled the nested hierarchy of precedence within the Mazu cult (Sangren 1988b, 1993), but now appear unlikely to alter or undermine it dramatically.

This brief excursion into developments in the Mazu cult has implications for broader questions of identity. It suggests that for many worshippers there is no perceived contradiction between Mazu's role as a symbol of Taiwanese identity, on the one hand, and her association with a wider sense of community that includes at least the whole of the *Min'nan* region, Taiwan, and those areas of Southeast Asia settled by overseas Chinese who brought their cult with them.

Moreover, the fluidity of the lines of association that constitute the Mazu cult parallels, in many respects, the fluidity of the situation with regard to Taiwanese identity. For example, one might see in the argument that Taiwan's Beigang Mazu is now the most authentic locus of the deity's tradition and power, a claim easily appended to a more pervasive sense of an emergent Taiwanese identity, one that separates itself not only from Mainlanders in Taiwan, but also from Fujian as the root or source of cultural identity. In fact, there are many in Taiwan who advocate such views. But although such sentiments are widely expressed in Taiwan, one should not overlook other crosscutting, complicating lines of possible social association.

In an article, Bosco (1992a) has argued persuasively for the emergence of a "Taiwanese popular culture." Bosco concludes that "the 'mainland fever' that broke out in Taiwan after 1987, has had the ironic effect of turning all residents of Taiwan, both Mainlanders and native Taiwanese, into 'Taiwanese'" (1992: 52). Some of the elements of his argument are worth summarizing: (1) The shifting balance of political power in Taiwan away from the Mainlanders has seen the prestige of *Min'nan* language increase rapidly, to the point where Mainlanders now actively attempt to learn and use it;[30] (2) Local and, in some cases, national politicians make it a point to embrace popular religion—a dramatic reversal of the policies of the 1950s and 1960s;[31] (3) In literature, magazines, and movies, there is a resurgent interest in the traditional, rural, and folk (especially Taiwanese). Moreover, this sometimes rather sentimental cultural trend is tacitly condoned by the government—again, a dramatic reversal of official anti-traditionalism of the 1950s and 1960s.

Bosco attributes these developments to a variety of circumstances—among them the particular ways in which the KMT managed to sustain its minority rule by encouraging a kind of factional local politics.

Most significant for this discussion, however, is Bosco's thesis that even Mainlanders are beginning to consider themselves "Taiwanese" and that Taiwanese resentment toward Mainlanders is diminishing:

> . . . the Taiwanese popular culture discussed in this paper unites what appear to be irreconcilable nationalisms. The election process with which the KMT institutionalized its rule has led in the late 1980s to greater pluralism and the emergence of localist, Taiwan-centered interests. The forced retirement of mainland representatives completes the Taiwanization process. The KMT will no longer be able to rely on mainlander-dominated national bodies for control. Democratization of Taiwan politics is easing the resentment against mainlanders and the KMT; it will no longer be necessary to argue that independence is necessary to remove the dominant mainlanders and achieve democracy. . . . Now, especially after the June 4 Incident, even the KMT argues that Taiwan is different economically, socially, and politically from the PRC and cannot be reunited with the mainland in the near future. (Bosco 1992a: 62)

Although Bosco may possess an overly sanguine view regarding how easily longstanding resentments can be forgotten (time will tell), his depiction of an emerging popular culture in Taiwan, one that transcends Mainlander/Taiwanese divisions, is important to keep in mind. And, although I would hesitate to predict the eventual dissolution of Mainlander/Taiwanese divisions in Taiwan, as Bosco points out, the palpable presence and *difference* of the People's Republic, perhaps even more so now that travel there is common, highlights the commonalities of interest and experience of the Taiwanese and the Mainlanders in Taiwan.

Bosco emphasizes the fact that "mainland fever" has highlighted the differences between Taiwan and the PRC, creating a context in which the Mainlanders and the Taiwanese find more common ground. Yet, one might also suppose that sustained increases in interaction between Fujian and Taiwan may increase the significance of shared *Min'nan* identity. One might hypothesize that some of the same trends Bosco documents—for example, the growing currency of the uniquely Taiwanese pastiche of local, Japanese, and Western elements in the formation of a new popular culture in Fujian—will begin to diminish differences. By the same token, Taiwanese investment in Fujian and the emerging development of new business practices there will, if present trends continue, also diminish differences. Finally, one should not forget that Taiwanese contributions to the reconstruction of temples in Fujian vests Taiwanese symbolically as well as financially there.

The cultural implications of this investment are worthy of further study. The nexus of ideas and values associated with the magical power of deities (*ling*) constitutes an implicit ideology of productive fertility associated with variously demarcated communities (Sangren 1987). Investment in this fertility, by means of contributions to communal temples, is widely thought to redound not only to the spiritual credit of the contributor, but also to her/his own productive power. Businesses (and politicians) are typically generous and conspicuous contributors, and their own prowess as productive, successful businessmen, or political leaders is thought (somewhat circularly) to derive in part from this generosity. When the Taiwanese contribute to mainland temples, they thus not only invest in the communities in which those temples are located, but also assert a linkage between their own individual productivity, power, and identity to those communities (Sangren 1991).

"Chineseness"

Even though the distinction between the Taiwanese and the Mainlanders was quite prominent during the period of my initial fieldwork in Taiwan, I never sensed any doubt whatsoever among my Taiwanese informants regarding their identity as Chinese. Neither has there been much doubt in the multifarious discussions among Chinese intellectuals, until recently, that some core values and social qualities characterize Chinese civilization (although there has been considerable difference of opinion about what these qualities are and which ought to be cherished and which abandoned). Linked to the question of Chinese identity are debates over categories like "modernity" and "tradition."

Modern Chinese intellectual history has been characterized by an engagement with broader intellectual movements—Darwinism and Marxism, modernism and democratic socialism, for example. For the past decade or so, Chinese intellectuals on both sides of the Strait have been quick to take up themes from avantgardist movements in the West—deconstruction, postmodernism, poststructuralism. Part of the appeal of such movements in China is that they promise affiliation with progressive political values, on the one hand, but are implicitly or explicitly anti-Marxist, on the other. Just as appealing is the aggressive denigration of the "West's" arrogant appropriation of reason, science, and objectivity. Adopting a postmodern, deconstructive stance, one can be at once modern (or even postmodern) and at the same time argue that Western-style modernity is, at best, old fashioned, and at worst, a sophisticated apology for Western colonialism filled with rhetorical ploys and arrogant self delusions.[32] One of the problems of such stances, however, is that any "representation" of Chineseness is, in principle, vulnerable to attack as "orientalist," or worse.[33]

I shall, rather unfashionably, align myself with a more conventional view; the many complexities of identify formation depicted here notwithstanding, it seems to me that assumptions of some level of encompassing Chineseness is no mere effect of the self-obsessed, utopian delusions of Chinese intellectuals or a paranoid projection of Western orientalism. In his paper, for example, Li Yih-yuan (1993) has produced a useful list of qualities that manifest among Chinese everywhere. I shall not attempt to produce a similar catalogue here; what constitutes a core of a "culture" moves quickly into issues of abstract theory and epistemology. However, in line with my opening argument about the "constructedness" of cultural identity, I will note that in the domain of popular religion images of administered cosmos—what Angela Zito has felicitously termed the imperium—play an important role, by most accounts, everywhere Chinese. In other words, assumptions of encompassing Chineseness is not limited to the self-conscious productions of traditional literati and modern intellectuals.

Heterogeneity within the conceptual space of China—*tianxia*, "all under heaven"—is the taken-for-granted model of traditional cosmos.[34] Even among arguably heterodox cults, some encompassing cultural sphere of unity is generally assumed (see, e.g., Bosco 1992b). For example, I was told by a Taiwanese practitioner of spirit possession (Hokkien *tangki*; Mandarin *jitong*) that only the gods registered in the Ming epic *Fengshen Yanyi* ought to be considered "orthodox" (*zheng*). *Fengshen Yanyi*, among many other things, is a story of dynastic legitimacy and transfer of the Mandate of Heaven. My point is that an image of Chinese cultural unity manifested most explicitly in the image of the imperium that can be found in folk stories, mythologies, performative genres is taken for granted in local traditions. However, one might account for the prominence of the "imperial metaphor" in Chinese popular culture (opinions diverge),[35] it registers at least this modicum of self-identification as Chinese.

Multiple Identities?

Lest I be misconstrued, I am *not* arguing that some objectively observable list of cultural attributes suffices to define Taiwanese as "Chinese." My point is, rather, that in many contexts, the Taiwanese represent themselves to themselves as Chinese. It is still an open question as to whether or not such representations will be superseded by emerging self-identifications as Taiwanese. By the same token, I would not predict that a sense of *Min'nan* identity is likely to supersede either conventionally defined Taiwanese (as opposed to Taiwanese Mainlander) identity or an emergent one that includes Mainlanders. I would argue, however, that some meaningful activities (for example, pilgrimages to Fujian temples, visits to puppet shops) already manifest the existence *in some contexts* of such an identity. The degree to which *Min'nan* language and historical connection become important in broader contexts will depend on future economic and political developments in both Taiwan and the mainland.

Tu Wei-ming (1991) has argued controversially that in many respects the works of intellectuals in the Chinese diaspora (which, in addition to those in the West, would also include those in Southeast Asia, Hong Kong, and Taiwan) have become central to the evolution of Chinese culture. Although there may be good reasons to dispute aspects of this argument, it seems to me to suggest a useful way to think about how best to conceive "Chineseness." "Chineseness," when considered in this fashion, is thus less a list of essential cultural attributes (although at some level, such attributes must be supposed to exist) than it is a sense of membership in an evolving transnational culture that encompasses quite divergent Chinese communities.

In this light, the state-dominated cultural hegemonies, totalized "bodies politic" (Anagnost 1994), that characterized imperial China and continue to hold ideological sway in official state policy in the PRC are no longer appropriate contexts for defining "Chineseness." That such state-centered ideologies of identity are increasingly perceived as anachronisms in Taiwan is a mark of the island's cultural pluralism and of its modernity. Consequently, at least in principle, one need not disavow membership in a wider transnational sphere of Chinese identity in order to locate oneself more specifically as "Taiwanese."

Much of the foregoing argument might be re-cast in a somewhat different light if we take note of an oxymoronic quality internal to the concept of "identity." In mathematics, equations express identities; that which appears on one side of the equation is the same thing—identical to—as that which appears on the other. The term connotes absolute discreetness, clear-cut boundaries, and things as such. Yet in social philosophy and cultural studies the notion that people are possessed of stable, singular "subjectivity" or identity is widely disparaged. Who and what we are is not the same all the time. More mundanely, one might say that people play different roles in different contexts. The same person might identify himself or herself as Taiwanese, Chinese, *Min'nan*, Quanzhou, male/female, a school teacher, from such and such a village, and so forth depending on context. Thus, a fervently pro-Taiwan independence shaman may also invoke a totalizing Chinese image of cosmos in ritual performances.

The demands of different "identities" can, obviously, pull people in different directions. I have Taiwanese friends who at once lament the narrow-minded dismissiveness of conservative Mainlander officials regarding the value of Taiwanese popular culture, and at the same time complain of a Taiwanese-style political correctness that in extreme forms insists on denying any relevant cultural connection with a greater Chinese tradition. The current political cum cultural scene in Taiwan requires as much politesse as does that in American universities. Moreover, the Taiwanese (and the Taiwanese Mainlanders) are acutely conscious of the sensitivity and fluidity of the current scene. Indeed, I would argue that the fluidity and sensitivity surrounding identity issues itself constitutes part of the emergent popular culture. Less "post-modern" than complex, Taiwan is a remarkably sophisticated, cosmopolitan society in which successful negotiation requires considerable self-consciousness and adeptness at handling identity issues in appropriate contexts.

Yet, modern and complex as Taiwan is, I suspect that in very different terms similar complexity has long been part of other Chinese scenes, especially for elites and urbanites. Some Western scholars are wont to see in Chinese epistemologies of person and body, state and society, modes of conception radically different from and often implicitly preferable to those attributed to "Western" thought (see, e.g., Zito and Barlow, eds., 1994; especially Zito 1994; Hay 1994; and Farquhar 1994). The general picture that emerges in such writings is that Chinese "selves" and "bodies" are less essentialized, less unified, more socially contextualized, and less associated with a totalizing, transcendental "subject."[36]

This is not the place to launch a lengthy discussion of such arguments. Indeed, the preceding comments regarding multiple identities might be viewed as consistent with them. Yet I believe that such interpretive efforts, highly enlightening in many important ways, sometimes overlook aspects of Chinese social life and culture that run counter to their desire to see in Chinese culture an antidote to perceived shortcomings of the West. With regard to the negotiation of the demands of multiple social roles ("subject positions," to use current jargon), for example, one might take note of the Chinese phrase *zuo ren*—literally, "to do/be a person." It is a very complementary thing to say of someone that he or she "*hen hui zuo ren*" ("is very good at being/doing a person"). Among the connotations associated with the uses of this phrase are the ability to be at once sensitive to the contextual demands and expectations of one's contemporaries, and to be able to locate one's own appropriate behaviors in the complex intersection of the desires of relevant others in varying social situations. In other words, role playing is more than simply applying a role associated with an identity to a context; to be an effective person, to *zuo ren*, is to "totalize" multiply intersecting roles and "identities" with sufficiently perceptive contextual sensitivity.

I bracket "totalize" in quotation marks because it is a refusal to totalize at the level of the self or person that is widely cited as a virtue of non-Western ethnopsychologies. My argument thus diverges from the overwhelming emphasis of recent scholarship. Although implicit notions of person and bodiliness in traditional Chinese medicine (Farquhar 1994) and art (Hay 1994) may convey a less transcendent, agentive, "present" sense of person than do their Western counterparts, practical consciousness does not. Chinese individuals may find the demands of competing social commitments—identities—difficult to negotiate successfully, but "multiple identities" should not be construed as implying lack of unity at the level of self-consciousness.

Conclusions

Invocation of the multiplicity of identifications available to residents of Taiwan and Fujian is admittedly a less than venturesome conclusion. I shall transgress my own stricture against prognostication here, however, and venture the less than muscular view that identifications as Taiwanese, Chinese, and *Min'nan* will all continue to constitute important elements of individuals' contextually practiced senses of themselves. More substantively, I agree with Bosco that there is emerging in Taiwan a popular culture in which both the Mainlanders and the Taiwanese participate, but I also believe that one should not overlook the possibilities both of renewed alienation between these groups and of the emergence of a sense of cross-Strait *Min'nan* identity. The latter development would depend very much on the continuation of (in light of events of 1993, perhaps "return to" would be a more appropriate phase) relatively liberal policies on both sides of the Strait.

I have digressed from issues restricted to religious dimensions of the practice and construction of identity, but I shall risk an additional, somewhat more venturesome, prediction in this regard: Identity issues that emerge in practical life—individual and collective—will all but inevitably manifest in religious activities. Future developments in the Mazu cult, for example, are likely to be useful indicators of the significance of a *Min'nan* identity that encompasses Mainland–Taiwan differences. By the same token, changing patterns of family development and gender roles are likely to manifest in shifting mythic emphases in extant cults and, perhaps, see the emergence of new ones.[37] The plasticity of Chinese religion as a language of individual and collective identity construction—what I prefer to term self-production—endows it with the ability to be both "traditional" and to adapt to changing social circumstances.

Afterword: *Minjian Xinyang*

Based on his participation in a remarkable succession of conferences on local religion on both sides of the Taiwan Strait during the past several years, Kenneth Dean (n.d.a) observes how various sponsors and participants—local, provincial, and national governments as well as academics—have sought to promote divergent readings of popular cults that align with equally various political interests. In a similar vein, I would like to add a brief commentary that derives in part, and quite impressionistically, from my own less extensive participation in Taiwanese academic life.

Academic study of "folk religion"—the Mazu cult most conspicuously—has clearly become an important arena of contestation over identity. Dean and Rubinstein remark upon some of the efforts of PRC officials to exploit the cult to advance their policy objectives toward Taiwan, for example. Here, I would like to comment on what is perhaps a more subtle manifestation of identity politics in the growing interest of Taiwanese academics in *minjian xinyang*. Over the course of the past decade, scholars trained in history, literature, and sinology have been drawn increasingly to the study of what is now widely termed *minjian xinyang* ("beliefs of the people"—what I am glossing as "folk religion"). The sinological skills that Taiwanese historians and sinologists bring to the study of local religion (for example, in foregrounding the role of liturgical traditions) have enriched enormously our understanding of how to locate local practices in wider historical and cultural terms. Writing as an anthropologist, I confess that the erudition of such scholars (and their Western counterparts) has challenged profoundly some of my own disciplinary premises, and compelled me to rethink aspects of my earlier work.

But also writing from the perspective of anthropology, I would add that the emergence of a broader interdisciplinary interest in folk religion is itself a phenomenon worthy of academic contemplation. My argument, in brief, is that the emergence of the category *minjian xinyang* as a relatively new object for sinological study is, in Taiwan, very much implicated in identity-politics issues.[38] During my recent sabbatical in Taiwan (August 1994–January 1995), I attended a number of lectures and meetings on *minjian xinyang*. A desire to describe and delineate a tradition unique to Taiwan clearly motivates the research of some scholars. By the same token, scholars who frame their work in terms that transcend Taiwanese/Mainland boundaries are frequently criticized for an inappropriate research focus, by implication one that legitimates the notion of a greater "cultural" China. The term itself, *wenhua zhongguo*, is often invoked as an unflattering epithet affiliated with the assumed pro-reunification sympathies of scholars who implicitly or explicitly assume some measure of Chinese cultural unity in their studies. The most explicit articulation of this critique is Hong and Murray's (1994) highly polemical review of Western anthropological studies of Taiwan.

I am reminded here of the debates in the 1970s, among Western anthropologists over the question of China's cultural unity. I reviewed these debates and outlined my own thoughts in an earlier publication (Sangren 1984). These debates were marked by principled disagreement over whether Chinese society is best comprehended as possessing some integrating, common cultural themes, on the one hand, or whether, crudely put, all culture is local. Current discussion in Taiwan seems to have taken little note of these earlier discussions, however. This oversight is unfortunate, at least from the viewpoint of anthropology, but understandable given the much more directly political implications of the issues in Taiwan.

What is unfortunate, lively disagreements notwithstanding, is that out of the preceding debates there seems to me to have emerged a vague consensus that "culture" is not discrete; it is not a quality that one either possesses or not. As I argued above, in principle, one need not be *either* Chinese *or* Taiwanese. Rather, culture is a complex patterning of processes that manifest *both* in characteristic patternings of social relations (including power relations, gender roles, political life) *and* representations (ideology, systems of symbols, beliefs). Commonalities in social relations and cultural life may be discernible among locales that also differ in significant respects. In other words, the concept of culture must embrace both the possibility of commonalities and of variations. Were this not the case, every individual, family, village, market system, and so on, could make a plausible claim to possess a unique culture.

Thus, while it is true that Taiwan does not represent Chinese culture as a whole in any typifying sense, neither does any other locale.

The typical Chinese village has no existence other than as an ideal type. Whether or not Taiwan should be considered *a* Chinese local culture (among many others) is both an empirical question and one whose answer in analytical abstraction would depend very much on definitions; how much variation does one wish to encompass.

Anthropology as a discipline is less helpful than it should be in clarifying such issues precisely because its central organizing concept—*culture*—is so vaguely and diversely conceived and deployed. To cite only two very influential exemplars, in the writings of Levi-Strauss and of Clifford Geertz one finds starkly divergent concepts of culture. For Levi-Strauss, there is only culture, parts of which are variously manifested in different times and places; for Clifford Geertz, there are only cultures, each uniquely comprised of systems of symbols constituting a world of meanings. For reasons too complex to explore fully here, in my view both of these widely influential definitions—often incoherently combined—have done more to obstruct than to advance anthropology.

In brief, Geertz's emphasis on uniqueness and insistence that cultures be understood—insofar as the literary skills of the ethnographer are capable of interpreting the terms and concepts of one culture into those of another culture—*in its own terms* produces philosophical confusions whose logical extensions into the excesses of postmodernistic posturings climaxed about a decade ago in American anthropology. One can agree with Geertz's eloquent defense of the importance of culture without conceding to his insistence that culture is text analyzable only as such.

Geertz's able defense of relativism aligns nicely with arguments of some currency in Taiwan to the effect that the only legitimate way to approach a culture is to allow it to speak for itself. It is imperative for anthropologists to take heed of both individual and collective self-representations. Indeed, it is one of the discipline's most worthy traditions to insist that culture is the organizing principle of social life and that social life cannot be understood as a mere consequence of individual or group maximizing strategies (usually with reference to self-evident "interests" related to equally self-evidently determined preferences or to "inclusive fitness" defined in terms of reproductive success).

Yet, anthropologists are also well advised to be wary of pitfalls in a relativist rhetorical appeal to what appear to be democratic values. A consequence of Geertz's relativism and insistence that culture is text is to ridicule analysis of culture's ideological operations; anthropological comparison of cultures and of anthropological analysis as cultural critique are delegitimated. The concept of culture epitomized by Geertz's work thus seems to me to entail two unfortunate consequences.

First, anthropology's ambition to constitute a science of social life and culture has eroded to the point where ethnography is defined as a literary account no longer testifying to the realities of its earlier objects—social and cultural life—but rather to the particularistic viewpoints of the ethnographer her- or himself. There is, in a very palpable and sometimes explicit sense, no longer an anthropological object—"culture"—to be studied, only the polymorphous assertions of multiple but transient identities competing to be heard.

Second, delegitimation of authorizing values (often denigrated as arrogant claims to objectivity and scientific status or as "master narratives") that might unite a community of scholars in debate within some common framing assumptions regarding what constitutes a relatively more convincing argument or analysis, implicitly legitimates a view to the effect not only that *all* voices have a right to be heard (who in good conscience would dispute such a value?), but also that *every* voice speaks *only* for itself and does so with absolute authority. This view, as I have argued elsewhere (Sangren 1987, 1988), is far less democratic than it seems. Any constructed representation of social and cultural realities ought to be framed in a way that is contestable; to be contestable, an argument must be coherent (even if this coherence encompasses contests and stochastic or random events and processes); in other words, it must make "truth claims" about realities that ultimately submit to the collective authority of logic and the judgment of other scholars.

The implicit construction of authority in ethnography, far from being the arrogation of unwarranted prerogatives as charged by critics like Clifford (1986), has long taken most of these framing premises for granted. The attempt to persuade through use of logic and evidence *assumes* the existence and prerogatives of an implicitly resistant, intelligent readership. Turning the tables on some postmodernistic posturings, the invocation of multivocality (unless it is coherently totalized at an encompassing level of analysis) and the apparently modest disavowal of truth-speaking authority (with regard to anything other than one's own point of view) seem to me to be far *less* democratic than is usually noticed. Such stances implicitly deny recognition of contestants' rights to dispute them by grounding authority not in the logical coherence or empirical evidence of an analysis (in *what* is said), but in the unassailable identity of the sayer (*who* says it).

In the domain of practical symbolic analysis, Levi-Strauss's work seems to me to have inspired more emulation of lasting ethnographic value. However, the notion that both the patterns of social life (e.g., kinship) and of symbolic representation (myth) manifest universal, underlying "structures of mind," like Geertz's vaguer notion of cultures as clusters of meaningful symbols, shortcircuits analysis of what ought to be the object of anthropology—*how* thought and action, symbol and practical life, constitute a dynamic process, a self-producing (hence, reproducing) system—by, in essence, *asserting* a unity of thought and action at the level of "culture" or of "underlying structure."

Given the historical and political circumstances that uniquely characterize Taiwan, analytical arbitrariness and definitional disputes over what constitutes "culture" have rhetorical force in practical politics not containable in the dispassionate terms of academic debate. In the domain of identity politics, *who* makes a claim carries a great deal of weight. Indeed, anthropological understanding entails noting how the distribution of power influences which definitions of culture prevail (e.g., those that would construct community around "Chineseness" or "Taiwaneseness"). Nonetheless, within the academe I believe that it might be appropriate for anthropologists interested in identity politics in Taiwan to keep in mind some of the larger issues surrounding the concept of culture. I shall not deny a measure of self-interest in this observation. It is obviously disadvantageous to a Western anthropologist interested in engaging in dialogue with one's Taiwanese colleagues when the value of one's work is measured solely or mainly by its unintended appropriability to either independence or unification agendas.

Crosscutting this sensitive issue is another—is Taiwanese (or, for that matter, Chinese) culture best approached through self-avowedly "native" categories of analysis, or is scholarship best advanced by an alliance between native sinological skill and "Western" "theory"? I bracket both "theory" and "Western" in quotation marks to indicate my discomfiture with both terms and by their association, at least in the present context. The notion that theory (relatively highly evaluated) and empirical scholarship (valued, but by implication less highly so) are separable endeavors, obviously evident in the statusing of intellectuals in Europe and the United States, is also evident in Taiwan's intellectual community. But so, too, is resistance to the status of theory—particularly "Western" theory.

The influence of various postcolonial and postmodernist critiques from the West is clear in an argument articulated with increasing frequency in Taiwan to the effect that Western theories should not guide Taiwanese scholars in orienting their own research efforts. Some of the reasons for this resistance are good ones, but in my view there is also some cause for concern that its energy is in danger of being misdirected. My discomfiture with the term "Western theory" is that it seems to take for granted the fact that because a mode of thinking has emerged in specific historical and cultural circumstances, its relevance for understanding differing historical and cultural circumstances is therefore compromised. This assumption, in my view quite problematic, aligns with the broadly based and directed critique of modernity that has excited not only Western, but also an emerging international avant-gardist intelligentsia.

Many compelling studies have shown that elements of what in an earlier era was considered "scientific" knowledge—for example, about race and intelligence or cultural evolution based on a notion of progress toward modern Western social forms—was in fact deeply compromised by assumptions both culturally specific and ideological. Moreover, such knowledge was and continues to be complicitous in the legitimation of institutions ranging from capitalism and imperialism to racism and patriarchy. But it is also important to emphasize that the critiques are themselves also products of historical and cultural circumstances.

Social Darwinist or racist "science" is not flawed because it is Western, it is flawed because it can be demonstrated *both* to be weak in logical and scientific terms *and* to have gained a certain credibility in particular historical circumstances precisely *because* its misrecognitions align with the interests of some sectors of society within the encompassing logic of social and cultural reproduction. In short, ethnocentrism may inform even the best-intentioned social analysis, and should be avoided. But by the same token, social analysis is legitimately disputed in logical and empirical terms on the basis of *what* is said, not by *who* says it.

"Theory," in other words, is not intrinsically Western or non-Western. The implicit values and assumptions that inform Chinese philosophy and Western social science, for example, differ in many ways. However, the premises that inform discussion and debate are not as radically incommensurable as sometimes assumed. My experience is that in the area of evaluating academic analyses, similarities are far more striking than differences. Evidence and logic have an ancient and finely honed tradition in Chinese philosophy, and I detect a welcome enthusiasm among my colleagues in Taiwan for ideas from the West that provide unexpected and convincing new ways to comprehend social life.

By the same token, I suspect that Taiwanese colleagues will continue to notice elements of ethnocentric biases in the writings of their Western counterparts, and Westerners may be more inclined to notice some of the culturally specific, taken-for-granted assumptions that inform the studies of Taiwanese scholars. Especially in the study of a phenomenon as rich and complex as folk religion, the conceptual tools at our disposal will undoubtedly require considerable refinement—refinement that I assume can only be advanced by a collaborative engagement between Taiwanese and non-Taiwanese researchers. In the event that I have not yet made my intentions sufficiently clear, I perceive some danger that the current efflorescence of multidisciplinary interest in folk religion and popular culture might be overwhelmed by the potential divisiveness of identity politics. However, I write in the more optimistic hope that the current convergence of circumstances can be harnessed to a productive reassessment of the ways we comprehend both the politics of cultural identity and, indeed, culture more generally.

Notes

1. This paper was first written for Conference on "Economy and Society in Southeastern China," East Asia Program, Cornell University, 2–3 October 1993. Subsequently, it was presented at the Department of Anthropology and Center for Chinese Studies, University of Washington, 9 October 1995 and at the Taiwan Studies Workshop, Harvard University, 23 October 1995.
2. I wholeheartedly agree with Dean and also Lagerwey (1987) that Taoist rituals, cosmologies, and the liturgical texts associated with them distill fundamental aspects of Chinese religion. Taoism, more than other textual traditions, sustains an intimate relation with local society. Yet I might emphasize more than do Lagerwey and Dean the dialectical relation between local social organization and Taoist text and ritual practice. While it is true that Taoism persists as an organizing, productive language whose practice plays an important role in producing and expressing local social solidarities, it is also worth emphasizing that this language is itself encompassed within the processes of local social production. Taoist ritual is, in short, an abstract and "alienated" language of production of local social institutions and "cultural subjects," but it is a language that is also a product of those same processes (Sangren 1987, 1991, 1994).
3. *Min'nan* ["south of the Min river"] is conventionally employed both as a linguistic category (as in "southern Min" dialect, Hokkien, the language spoken in the Zhang-Quan region of southern Fujian) and as a cultural gloss (numerous Chinese publications employ the term *Min'nan ren* to refer to the common historical and cultural attributes shared by *Min'nan* speakers). The choice of such terms against others (for example *Taiwan ren* or *Taiwan hua*) can be an explicit or implicit indicator of the political leanings of their users. One might note that Mandarin, too, is commonly glossed in Taiwan as *guoyu* ("national language"), in mainland China as *putonghua* ("common or standard speech"), and in Singapore as *huayu* ("Chinese"—implicitly a category indicating cultural as opposed to national identity). Obviously, these terms are intentional and reflect, among other factors, differing identity politics.
4. For example, as I revise this essay (September 1994), the PRC government's apparently successful effort to prevent President Li Denghui's visit to the

Asia Games in Yokohama seems to have produced all but unanimous indignation in Taiwan. Both the KMT and its opposition (including Taiwan independence advocates) believe that Taiwan's economic and political achievements have earned it a more prominent voice in international politics. If the Beijing government had hoped to take any advantage of division in Taiwan, its current policies seem remarkably counterproductive.

5. It is a mark of the rapidity of changing consciousnesses in Taiwan that there has been a lively public debate over terms appropriate to designate aboriginal peoples. When I conducted fieldwork in the mid-1970s in a Zhangzhou-dominated area bordering mountainous Fuxing *xiang*, inhabited mainly by Atayal aboriginal people, the term *fanzi* (H. *hoan-a*)—"barbarian"—was the most common term used by Daxi people; *shandi ren* ("mountain people") and the more scholarly *gaoshan zu* ("mountain tribes") were employed both by more sensitive Han and by aboriginal people themselves. In recent years, not only *fanzi*, but even *shandi ren*, have become politically incorrect. Even the term *shaoshu minzu* ("minority peoples") is somewhat problematic because to some it suggests less than full participation and rights. A widely publicized dispute involving government officials, aborigine rights advocates, and prominent anthropologists focused on what the proper term ought to be. Two candidates are *yuanzhumin* ("original inhabitants") and *xianzhumin* ("prior inhabitants"). It is also noteworthy that aboriginal ancestry has transformed from a shameful quality, to be denied if possible, to a desirable quality sometimes claimed on rather implausible bases. (In the mid-1970s in the Daxi area, for example, descendants of Taiwan's "plains aborigines" [*pingpu zu*] steadfastly claimed mainland Han origins, even though they lived in villages that clearly identified them as *pingpu zu*.) In part, this transformation is linked to the claims that Taiwanese make against mainlanders to be Taiwan's rightful proprietors. Obviously, aboriginal peoples can make similar claims against Han Taiwanese. Moreover, if Han Taiwanese can claim some aboriginal ancestry, they can hope to make common cause with aboriginal people against Mainlanders. But this scenario requires a radical shift in emphasis in the narrative of Taiwanese history and settlement. In this standard narrative—seen, for example, in the story of the spread of the Mazu cult to Taiwan—Taiwan's settlement from the *Min'nan* region is a continuation of a kind of manifest destiny of the spread of Han civilization. The taken-for-granted superiority of Han culture accounts in such narratives for the displacement of aborigines, who were forced into the less productive mountainous areas. There is little acknowledgment in this standard narrative for assimilation of aboriginal peoples. The situation has altered to the point that some Taiwanese resist being termed Chinese or Han, reserving such terms for derogation of Mainlanders.
6. Most Taiwanese Hakka also speak Hokkien. Although I lack statistical data confirming the point, it is undoubtedly the case that relatively few Hokkien speakers attain equivalent facility in Hakka.
7. Dean's recent (1993a) and forthcoming (n.d.a and n.d.b) work is instructive in this regard. His research project on Putian and Xinghua promises to

provide a thoroughgoing map of the embedded layers of social identities and their changes through time at a scale of analysis and in detail unavailable to us in previous studies. See Sangren (1987) for a smaller-scaled local study of the marketing system focusing on Daxi, Taiwan.

8. A recurrent theme in several of the papers of a recently (1994) held AAS symposium on religion and ritual in southeast China is that in such activities as ceremonial processions (in which gods, symbols of variously defined social groups are paraded on *youjing* ["inspecting the boundaries"] tours of their "domains"), encounters between gods frequently entail rather tense performative displays and both "mock" and, occasionally, real battles between their respective advocates. In the rhetoric of disputes over status in a celestial bureaucracy, what seems to be at sake in these encounters are issues of community boundaries and precedence. A similar sort of contestation over community construction is evident in the ongoing disputes among rival centers of the Mazu cult (Sangren 1988b, 1993).
9. The fact that local and national politicians increasingly seek to gain the "approval" of local gods (by means of conspicuous contributions, obeisances, and advocacy at higher levels of government) is one of the means by which local communities acquire legitimacy and recognition from the government. The secular trend in Taiwan has been a shift from wary and even hostile relations between central power holders and local religious organizations in the mid-1960s toward a more mutually supportive relationship. I view this as a measure of increasing cultural and political integration in Taiwan, or, in Skinner's (1971) terms, of increasing local-system "openness."
10. Although it is the impact of travel to Fujian that engages us here, one should not overlook the perhaps even more profound effects of travel to Japan and the United States. It is procedurally easier for Taiwanese to travel abroad now, and many more can now afford to do so. Since the late 1980s Taiwanese have been allowed to attend college abroad, whereas earlier such travel was allowed only for graduate school (and among males, only to those whose military-services obligations had been completed). The competitiveness of Taiwan's college-entrance examinations recommends this as a strategy to many. I do not have statistics at hand, but those familiar with the Taiwan scene will concur that most middle-class Taiwanese have had at least some experience abroad, and many are foreign-educated. The impact of these experiences is evident, for example, in the changing consumer tastes of Taiwan (creating the demand that has made McDonald's, Pizza Hut, and Seven-Eleven so successful), and in the desire among many Taiwanese professionals to reproduce aspects of Western middle-class life-styles. In short, Taiwan is rapidly becoming a very cosmopolitan society, not only in life-style, but also in political values and consciousness.
11. Taiwan government statistics cited in *Far Eastern Economic Review*, 15 September, 1994: 63.
12. I made a brief (one-week) visit to Fujian in August 1992 when I visited the Mazu temple in Meizhou and several temples in Xiamen and Quanzhou. I made a second visit during August 1994. I have been a frequent

intermittent visitor to Taiwan (once every year or so) since my dissertation fieldwork in the mid-1970s. My recent studies in Taiwan have moved from an early focus on local religion (Sangren 1987) to pilgrimages and the Mazu cult (1988b, 1991, 1993). Most recently I have been engaged in analysis of myths and gender, particularly surrounding the cult of the god Taizi, a personage better known as Nuozha from the Ming epic *Fengshen Yanyi*.

13. See, for example, Murray and Hong (1991, 1994) and Hong (1994).
14. The prevalence of this usage is probably most directly attributable to Clifford (1986), whose intention is to emphasize the constructed quality of identities and cultures. However, Clifford also intends to associate social production here with literary or linguistic production. Clifford's work epitomizes a move toward linguistic models of culture that has become extremely influential. I develop an extensive critique of some of the limitations of an overly constructivist view of culture (particularly of poststructuralist and interpretivist treatments) in Sangren 1988a, n.d.c, n.d.d. One of the most salient of these limitations is that agency and intentional action tend to be elided by the emphasis on the semantic givenness of culturally imagined realities.
15. See also Bosco (1992a), who takes a broadly similar position.
16. I quote from the title of an essay I wrote (Sangren 1988b) on the uses of historical narrative in rivalry for cult status in the Mazu cult on Taiwan.
17. Cultural identity thus has ideological qualities, especially to the degree that it is conceived as an essential or inherent quality of persons and collectivities, and to the degree that its genesis in discursive cum social practices is not explicitly recognized.
18. To be precise, one should employ Skinner's (especially 1964–65, 1977, 1980) regional-analysis framework here; the "natural" systems of market towns and cities define social space more relevantly than do administrative district boundaries. I follow local convention, however, in borrowing administratively derived glosses; local place names do not usually correspond directly to Skinnerian local systems and regions.
19. It is noteworthy that the fact that Hakka speak their own language further complicates clear definitions of Taiwaneseness. The history of Hakka settlement parallels that from the Zhang-Quan region; their claims to historical precedent are just as strong as are those of "Taiwanese" speakers. Consequently, for example, demands to make Taiwanese an official language have caused some Hakka advocates to respond with a demand for similar status for the Hakka language.
20. This pattern of native place affiliation or *tong*ism is a familiar one to students Chinese society. Chinese communities in Southeast Asian cities, for example, typically organize themselves around nested hierarchies of native-place associations (Crissman 1967); and native-place ties formed the bases for occupational specialization in the economic organization of Chinese cities (Skinner 1976).
21. A rather unusual statement for the owner of a shop selling religious articles . . . One should, of course, avoid placing too much weight on its literal veracity. It may be that this spontaneously offered admission was

motivated in part by the fact that I am a Westerner—that is, the tourist may have articulated what he assumed would be a stance more acceptable to me. But even granting such a possibility, his subsequent elaborations regarding cultural value seem to me compatible with a more conventional religiosity. Indeed, it is my impression that many Taiwanese engage in religious rituals both because they believe in the supernatural efficacy of gods and, quite self-consciously, because participation is valued for its own sake as a manifestation of social solidarity and tradition. The same person might cite either motive, depending on circumstances. In sum, although sincerity of belief is a value frequently emphasized in Taiwanese religion, other motives are also considered quite proper and complementary aspect of participation. Taiwanese communal rituals, especially processions of territorial-cult gods, thus combine the sentiments of celebrations of community (e.g., Fourth of July, Memorial Day parades) and of god (individual worship, going to church).

22. When I visited him in 1995, he informed me that he had, indeed, returned to purchase the robe, which he subsequently sold upon his return to Taiwan.
23. Fuzhou also seems to possess considerable repute in Taiwan for its image carving. I interviewed a Tainan image carver who claimed high status on the basis of Fuzhou lineage in 1986. More recently, several participants at a Taipei exhibition of "Buddhist articles" claimed Fuzhou lineage (*Fojiao Shenghuo Wenwu ji Yipin Zhan*, pamphlet 1994).
24. I am informed by Wang Ch'iu-kuei (personal communication, 1994) that even earlier visits were arranged by the Xinghua *Tongxian hui* (native-place association) of Taipei and by a Mazu temple in Suao (a fishing port in Ilan county). Professor Wang is also of the opinion that the accessibility of Fujian has rapidly devalued the prestige of these newly acquired images. For a thorough study of Dajia Mazu, see Huang (1994). See also Chen (1984).
25. I spoke at some length with the proprietor of a shop in Meizhou selling religious articles and other souvenirs to pilgrims and tourists. The proprietor noted that the numbers of Taiwan visitors has fallen off considerably since the late 1980s, a fact he attributes to the fact that for many, one trip is enough. Having seen the famous Meizhou Mazu, many visitors may feel less inclined to make a second trip. I would add that the prestige of having visited Meizhou Mazu is not increased significantly by multiple visits, and now so many Taiwanese have visited the mainland, that the novelty has eroded.
26. This circumstance may be changing with a liberalization of control over religious activities, at least in some areas, on the mainland.
27. Anagnost (1987, 1994) argues that in official rhetoric the category "feudal superstition" serves an ideological purpose; it defines a demonized, devalued "other" against which the state can imagine itself the empowered, legitimate subject—producer of value, order, and modernity against backwardness, chaos, and decadence. In this regard, present-day socialist rhetoric bears some affinities with elite discourses in imperial times that, in altogether different terms, deemed magic and belief in spirits delusory and

not conducive to a properly ordered society. One should keep in mind, however, that the polarizations apparent in representative rhetoric should not be mistaken for clearcut bifurcations of "modern" versus "traditional" aspects of social life, on the one hand, or of "elite" versus "popular" culture, on the other (Sangren 1984).

28. Since I began drafting this paper, I attended a conference in Putian (11–15 August, 1994) organized by Kenneth Dean and Zheng Zhenman under the auspices of the Xinghua Local History Research Institute. At the conference, Dean reported on his recent visit to a Singapore branch of the "Three-in-One" religion he has been studying. Of immediate interest here is the fact that overseas Chinese are financing elaborate ritual performances in Fujian. Among their motives are the obvious credit such sponsorship brings them in both in the Fujian locales where the rituals are performed and in their diasporic communities. In addition, overseas Chinese can obtain higher quality, more elaborate performances, for less expense, employing more expert, higher-statused ritual specialists in Fujian. (My thanks to Ken Dean for alerting me to these circumstances—circumstances that obviously enhance the prestige of tradition in Fujian.)
29. A potential, at least, for rivalry exists between the Meizhou temple, the site of Mazu's apotheosis conventionally recognized as the cult's ancestral temple (*zumiao*), and a nearby temple at Mazu's reputed birthplace in Gangli Cun opposite Meizhou island on the Putian coast. Taiwan pilgrims usually visit both temples.
30. One might add that among urban, middle-class ethnic Taiwanese Mandarin has become the dominant language, but here too there is also an effort to recapture mastery of *Min'nan*. Not all of the college students studying Taiwanese are Mainlanders!
31. To cite just the most dramatic example, in 1980 President Chiang Ching-kuo made a highly publicized presentation of a Mazu image to the Qiaotian *Gong* (Cohen 1991:132; cited also in Bosco 1992a:58). Such efforts to reclaim local religion as "Chinese" (as opposed to exclusively Taiwanese), on the one hand, are perceived as politically motivated, but, on the other hand, this perception does not necessarily mean that they are unappreciated by Taiwanese.
32. Political power holders, in contrast to intellectuals, insist on China's difference on other grounds, usually invoking a more conservative "tradition"—a kind of paternalistic elevation of order as the key to Chinese culture and aspirations. The recent statements of Singapore officials regarding the shortcomings of what they characterize as self-indulgent Western culture and values, for example, seem to me quite similar to attitudes evident in Taiwan and PRC official circles. Such attitudes might be characterized as traditionalistic rather than traditional; it is less the case that a traditional social order is defended for its own sake than that order (and authority) are viewed as necessary for "modernization." In an inversion of Western avantgardist polemics against institutionalized reason—Foucault's normalizing discourses—Chinese governments characterize Western culture as dissimulating, entropic. Clearly, such self-legitimating demonizing of "otherness" is not limited to Western "orientalism."

33. I have in mind here, of course, Said's (1978) extremely influential critique. I am much in sympathy with arguments disputing elements of Said's advanced by Ahmad (1992).
34. I survey arguments regarding the unity or diversity of Chinese culture in Sangren 1984. One of the key points in my argument is that one can see underlying similarity in the symbolic and practical means by which difference is produced and expressed. The most obvious examples are patriliny (which divides essential categories of kin and nonkin) and territorial cults (one is an insider or an outsider).
35. See Sangren (1987) for my own account and for some of its differences with those of others.
36. A more useful aspect of the widespread critique of Western "Cartesian" assumptions is that it has spawned an interest in how linguistically constructed "subject positions" imply unequal power relations. Foucault is widely credited with this insight, but his writing seems to me to muddle more than advance understanding of the relations between "power" and the "subject" (Sangren n.d.a). Unlike Foucault and his emulators, it seems to me that the solution is not so much emphasizing how power circulates "between" "located" selves as it is who ends up empowered when lines of collective and individual identities are represented in certain ways. An additional difficulty surrounds the conflations of language and practical action within the omnibus category "discourse," as though the epistemological shortcomings of a misleading dichotomy could be transcended merely by inventing a term that claims to do so.
37. One of the characteristics of Taiwanese representations of magical power is that gods frequently manifest their power (*xianling*). The frequency of such manifestations and the variety of techniques by means of which supernatural power communicates to human beings means that transcendent order—heaven—is not static and changeless (although at a "metatranscendent" level, it is) (Sangren 1987). In short, the language of Taiwanese mythical representation is a very plastic one.
38. The phenomenon of *nihonjinron* in Japan seems broadly similar in this regard.

References

Ahmad, Aijaz. 1992. *In Theory: Class, Nations, Literatures*. New York: Verso.

Anagnost, Ann S. 1987. *Politics and Magic in Contemporary China*. Modern China 13: 40–61.

——. 1994. "The Politicized Body," in Zito and Barlow, eds., *Body, Subject & Power in China*. Chicago: University of Chicago Press. Pp. 131–156.

Baity, Philip C. 1975. *Religion in a Chinese Town*. Asian Folklore and Social Life Monographs, No. 64. Taipei: Orient Cultural Service.

Bosco, Joseph. 1992a. "The Emergence of a Taiwanese Popular Culture," in *American Journal of Chinese Studies* 1: 51–64.

——. 1992b. "Yiguan Dao: 'Heterodoxy' and Popular Religion in Taiwan," in Murray A. Rubinstein, ed., *The Other Taiwan, 1945 to the Present*. New York: M. E. Sharpe. Pp. 423–444.

Chen Min-hwei. 1984. A Study of Legend Changes in the Ma Tsu Cult of Taiwan: Status, Competition, and Popularity. Masters Thesis, Department of Folklore. Indiana University.

Clifford, James. 1986. "Introduction: Partial Truths," in James Clifford and George E. Marcus, eds., *Writing Culture: The Poetics and Politics of Ethnography*. Berkeley: University of California Press. Pp. 1–26.

Cohen, Myron L. 1991. "Being Chinese: The Peripheralization of Traditional Identity," in *Daedalus* 120(2): 113–134.

Crissman, Lawrence. 1967. "The Segmentary Structure of Urban Overseas Chinese Communities," in *Man* 2: 185–204.

Dean, Kenneth. 1993a. *Taoist Ritual and Popular Cults of Southeast China*. Princeton: Princeton University Press.

——. 1993b. Conferences of the Gods: Popular Cults Across the Taiwan Straits. Paper presented at Conference on Economy and Society in Southeastern China, East Asia Program, Cornell University, 2–3 October 1993.

——. n.d.a. Irrigation and Individuation in the Putian Plains. Manuscript.

——. n.d.b. Multiplicity and Individuation: The Temple Network of the Three in One in Putian and Xianyou. Manuscript.

DeGlopper, Donald R. 1974. "Religion and Ritual in Lukang," in Arthur P. Wolf, ed., *Religion and Ritual in Chinese Society*. Stanford: Stanford University Press. Pp. 43–70.

Farquhar, Judith. 1994. "Multiplicity, Point of View, and Responsibility in Traditional Healing," in Zito and Barlow, eds., *Body, Subject & Power in China*. Chicago: University of Chicago Press. Pp. 78–102.

Feuchtwang, Stephan. 1974. "Domestic and Communal Worship in Taiwan," in Arthur P. Wolf, ed., *Religion and Ritual in Chinese Society*. Stanford: Stanford University Press. Pp. 105–130.

Fojiao Shenghuo Wenwu ji Yipin Zhan ["Buddhist-Life Cultural and Artistic Objects Exhibition"] 1994 [catalogue of exhibitors].

Harrell, Stevan. 1982. *Ploughshare Village: Culture and Context in Taiwan*. Seattle: University of Washington Press.

Hay, John. 1994. "The Body Invisible in Chinese Art?" in Zito and Barlow, eds., *Body, Subject & Power in China*. Chicago: University of Chicago Press. Pp. 42–77.

Hong, Keelung. 1994. "Experiences of Being a 'Native': Observing Anthropology," in *Anthropology Today* 10, 3 (June): 6–9.

Huang Meiying. 1994. *Taiwan Mazu ti Xianhuo yu Yishi [Ceremonies and Incense Fire of Taiwan's Mazu]*. Taipei: Zili Wanbaoshe Wenhua Chubanbu.

Knapp, Ronald G. 1976. "Chinese Frontier Settlement in Taiwan," in *Annals of the Association of American Geographers* 66: 43–59.

Lagerwey, John. 1987. *Taoist Ritual in Chinese Society and History*. New York: Macmillan.

Lamley, Harry J. 1981. "Subethnic Rivalry in the Ch'ing Period," in Emily Martin Ahern and Hill Gates, eds., *The Anthropology of Taiwanese Society*, Stanford: Stanford University Press. Pp. 282–318.

Li Yih-yuan [Li Yi-yuan]. 1993. "Cong Minjian Wenhua Kan Wenhua Zhongguo" (Viewing the National Culture of China from the Perspective of

Popular Culture)." Paper presented at Conference on Wenhua Zhongguo Zhanwang: Linian yu Shiji [Hopes for the Development of China's National Culture: Ideals and Realities], Chinese University of Hong Kong, Zu Shao Tang, 10–12 March.

Murray, Stephen O. and Keelung Hong. 1991. "American Anthropologists Looking Through Taiwanese Culture," *Dialectical Anthropology* 16: 273–299.

——. 1994. *Taiwanese Culture, Taiwanese Society: A Critical Review of Social Science Research Done on Taiwan*. New York: University Press of America.

Rubinstein, Murray A. 1993. Fujian's God's/Taiwan's Gods: Reconstructing the Minnan Religious Matrix. Paper presented at Conference on Economy and Society in Southeastern China, East Asia Program, Cornell University, 2–3 October 1993.

Said, Edward W. 1978. *Orientalism*. New York: Random House.

Sangren, P. Steven. 1984. "Great Tradition and Little Traditions Reconsidered: The Question of Cultural Integration in China," in *Journal of Chinese Studies* 1: 1–24.

——. 1987. *History and Magical Power in a Chinese Community*. Stanford: Stanford University Press.

——. 1988a. "History and the Rhetoric of Legitimacy: The Ma Tsu Cult of Taiwan," in *Comparative Studies in Society and History* 30: 674–697.

——. 1988b. "Rhetoric and the Authority of Ethnography: 'Post-Modernism' and the Social Reproduction of Texts," in *Current Anthropology* 29: 405–435.

——. 1991. "Dialectics of Alienation: Individuals and Collectivities in Chinese Religion," in *Man* 26: 67–86.

——. 1993. "Power and Transcendence in the Ma Tsu Pilgrimages of Taiwan," in *American Ethnologist* 20: 264–282.

——. n.d.a. The Vicissitudes of Power: Chinese, Foucaultian, and Anthropological Conceptions of Power (forthcoming in Cultural Anthropology).

——. n.d.b. "Gods and Familial Relations: No-cha, Miao-shan, and Mu-lien," forthcoming in *Gods and Society in Late Imperial and Modern China* (tentative title), Meir Shahar and Robert Weller, eds.

——. n.d.c. Post-Structuralist and Interpretivist Approaches to Subjectivity and Personhood: A Critique.

——. n.d.d. Myths of Self Production: With Special Attention to the Story of Na-cha from the Feng-shen Yen-i.

Schipper, Kristofer M. 1977. "Neighborhood Cult Associations in Traditional Tainan," in G. William Skinner, ed., *The City in Late Imperial China*. Stanford: Stanford University Press. Pp. 651–676.

Skinner, G. William. 1964–65. "Marketing and Social Structure in Rural China, Parts I, II, and III," in *Journal of Asian Studies* 24: 3–43, 195–228, and 363–399 (reissued as a pamphlet in 1974: Ann Arbor: Association for Asian Studies).

——. 1971. "Chinese Peasants and the Closed Community: An Open and Shut Case," in *Comparative Studies in Society and History* 13: 270–281.

——. 1976. "Mobility Strategies in Late Imperial China: A Regional Systems Analysis," in Carol Smith, ed., *Regional Analysis I: Economic Systems*. New York: Academic Press. Pp. 327–364.

——. 1977. "Cities and the Hierarchy of Local Systems," in G. W. Skinner, ed., *The City in Late Imperial China*. Stanford: Stanford University Press. Pp. 273–352.

——. 1980. Marketing Systems and Regional Economies: Their Structure and Development. Paper prepared for the Symposium on Social and Economic History in China from the Song Dynasty to 1900, Chinese Academy of Social Sciences, Beijing, 26 October–1 November 1980.

Tu, Wei-ming. 1991. "Cultural China: The Periphery as the Center," in *Daedalus* 120: 1–32.

Weller, Robert P. 1987. *Unities and Diversities in Chinese Religion*. Seattle: University of Washington Press.

Zito, Angela. 1994. "Silk and Skin: Significant Boundaries," in Zito and Barlow, eds., *Body, Subject & Power in China*. Chicago: University of Chicago Press. Pp. 103–130.

Zito, Angela, and Tani E. Barlow, eds. 1994. *Body, Subject & Power in China*. Chicago: University of Chicago Press.

Index